Tourism and Prosperity
in Miao Land

Anthropology of Tourism:
Heritage, Mobility, and Society

Series Editors

Michael A. Di Giovine (West Chester University of Pennsylvania)
Noel B. Salazar (University of Leuven)

Mission Statement

The Anthropology of Tourism: Heritage, Mobility, and Society series provides anthropologists and others in the social sciences and humanities with cutting-edge and engaging research on the culture(s) of tourism. This series embraces anthropology's holistic and comprehensive approach to scholarship, and is sensitive to the complex diversity of human expression. Books in this series particularly examine tourism's relationship with cultural heritage and mobility and its impact on society. Contributions are transdisciplinary in nature, and either look at a particular country, region, or population, or take a more global approach. Including monographs and edited collections, this series is a valuable resource to scholars and students alike who are interested in the various manifestations of tourism and its role as the world's largest and fastest-growing source of socio-cultural and economic activity.

Advisory Board Members

Quetzil Castañeda, Saskia Cousin, Jackie Feldman, Nelson H. H. Graburn,
Jafar Jafari, Tom Selwyn, Valene Smith, Amanda Stronza,
Hazel Tucker, and Shinji Yamashita

Books in Series

Alternative Tourism in Budapest: Class, Culture, and Identity in a Postsocialist City, by Susan E. Hill
Tourism and Prosperity in Miao Land: Power and Inequality in Rural Ethnic China, by Xianghong Feng

Tourism and Prosperity in Miao Land

Power and Inequality in Rural Ethnic China

Xianghong Feng

LEXINGTON BOOKS
Lanham • Boulder • New York • London

Published by Lexington Books
An imprint of The Rowman & Littlefield Publishing Group, Inc.
4501 Forbes Boulevard, Suite 200, Lanham, Maryland 20706
www.rowman.com

Unit A, Whitacre Mews, 26-34 Stannary Street, London SE11 4AB

British Library Cataloguing in Publication Information Available

Library of Congress Cataloging-in-Publication Data Available

ISBN 978-1-4985-0995-4 (cloth : alk. paper)
ISBN 978-1-4985-0996-1 (electronic)

To my grandparents,

who raised me in the countryside until my school age.

They not only provided me a happy childhood,

but also educated me in their own way.

Contents

Acknowledgments

As always, I am most grateful to you, the people of Fenghuang. For many of you, we have been part of each other's lives for the past fifteen years. You have opened your doors, and shared with me your life, your world, and your views about them, which not only provided invaluable research material, but also enriched me as a person. You are my inspirations for work and for life.

I enjoyed living among you in Fenghuang. Many times, I forgot that I was there as a researcher. Once in a while, tourists approached me in the street to ask for directions or where to buy the best ginger candy and silver jewelry, mistaking me as a local. I was flattered when being told, "if you don't speak, we can barely tell you are not from here."

I cherish all the friendship and care you have all shown to me. I remember the night before my leaving Longcun village, San Ge stayed up late after a long day of work, to make a jar of Miao-style preserved red peppers for me to take; I remember the smoked pork Jing's family had always saved for me, expecting my annual return; I remember the moment when watching Liang walking away in the scorching July sun with her infant son on her back, after dropping me a new woven basket as a gift, which she just shopped for during a hot noon; I remember the countless times that Xia took me to eat at her home and the homes of her parents and her sister, with all those delicious dishes which warmed my stomach, and more importantly, with their companionship which warmed my heart. Even if one day I stop writing about Fenghuang, I will still return to you all. It is a habit.

I acknowledge with warm thanks the valuable support of the government officials at the state, the prefecture, and the county levels, including Mr. Yu-gang Shi at the State Ethnic Affairs Committee of China, the field assistance provided by Mr. Jun Shi at the People's Government of Xiangxi Tujia-Miao Ethnic Groups Autonomous Prefecture, Ms. Rufen Tian and Ms. Xuefen

Teng at the county Ethnic Affairs Bureau, and Mr. Yao Ma at the county Statistics Bureau. Ms. Rufen Tian treated me as her daughter, and went out of her way to make my life easier, and she is forever my Aunt Tian; Mr. Yao Ma always took me to the government canteen for lunch whenever I visited his office, and even tried to loan me his meal card so that I could go there to dine on my own.

I am deeply indebted to my Ph.D. advisor Dr. John H. Bodley. Despite the fact that I had little background in anthropology at the time, you took me under your wing after I had left China for the first time in my life to attend Washington State University's doctoral program in anthropology. You inspired me to become an anthropologist, a researcher, and a university professor, a path that I had not planned for myself, but I will never trade it for any other. I thank Dr. John Young who encouraged me to start working on this book project, and I also thank him for his insightful comments on my book proposal.

The book is based on the research I have conducted since I started my position as an assistant professor of anthropology at Eastern Michigan University. It has been supported by several internal fellowships and awards, which largely funded my summer field trips, and provided me release time for writing during regular semesters. I thank my colleagues at Eastern Michigan University, especially the faculty at the Department of Sociology, Anthropology, and Criminology who have been the most supportive to their junior faculty, and I feel fortunate to be part of this department.

Earlier versions of parts of this book have appeared previously in "Who Are the 'Hosts'?: Village Tours in Fenghuang, China," *Human Organization* (2012) 71(4); "Space, Power, and Tourism: Notes from the Field in Rural Ethnic China," *Anthropology News* (November 2012) 53(9); "Women's Work, Men's Work: Gender and Tourism among the Miao in Rural China," *Anthropology of Work Review* (2013) 34(1); "From Labor to Capital: Tourism and the Poverty of Resources in Rural Ethnic China," *Urban Anthropology and Studies of Cultural Systems and World Economic Development* (2012) 41[2, 3, 4]; "Protesting Power: Everyday Resistance in a Touristic Chinese Miao Village," *Journal of Tourism and Cultural Change* (2015) 13(3). A Chinese version of Chapter 7 was published previously in "竞争与(不)平等：湖南凤凰苗寨游家庭餐馆的个案研究 [Competition and (In)equality: A Case Study of the Family Restaurants in a Touristic Miao Village in Fenghuang County of Hunan, China]," *旅游学刊* [*Tourism Tribune*] (2016) 31(3), and Chapter 7 is revised from that. I thank the publishers for permissions to use these materials.

Additional thanks to Ms. Amy King, my former editor at Lexington Books, who reached out to me and worked out a book contract, as well as provided valuable anonymous external reviewers' comments to my first draft before she took another position elsewhere, and to Ms. Kasey Beduhn, my current editor at Lexington Books. This book is one of the first volumes to appear in the series Anthropology of Tourism: Heritage, Mobility, and Society, and I thank the series editors Dr. Michael A. Di Giovine and Dr. Noel B. Salazar for their valuable feedback to the drafts of this manuscript. I also thank the external reviewer for the insightful comments.

I want to thank Mr. Robert Kubis. Thank you for your magic to shape me into a dancer and help me discover the other passion that I never saw myself. After long hours of writing, spinning on the ballroom floor has brought me the most joy and set me free.

My final thanks are to my family and friends in China. Without their moral support, I could never dream of coming this far in my career. I thank my parents for allowing me, their only child, to go so far away from them. Tengming Lei (雷腾铭), thank you for being one of the most important people in my life; without you, I doubt I had the courage to come to the United States alone to pursue a doctoral degree, let alone finish it.

My special thanks is to Dr. Mark Hill, I always count on you as the first reader for any of my writings, and I thank you for your extreme patience and insightful comments as well as your help with the map illustrations. Most importantly, I thank you for always being there for me. You are my safe harbor.

Introduction

Into the Field

On July 6, 2014, four days after my arrival in Beijing from Michigan, San Ge[1] called with the news. He told me how he had renovated his old mud-brick house in the village of Longcun[2] and used it to open a restaurant. He said he had done most of the renovation himself, losing 20 pounds in the process. He had also learned how to make rice liquor so he could sell it at his new restaurant.

"Send me pictures of the restaurant," I said.

"Why don't you come and see it?" he replied, knowing that I had no plan to visit for fieldwork that summer. He continued, "You come and help me. I am busy in the kitchen, and I need a hand to sell the liquor."

Tempted, I imagined myself sitting by the restaurant door selling rice liquor to tourists. "Let me think about it," I said as the idea began to take hold.

"Make up your mind quickly then. The sooner, the better. I do need your help. And this time you can stay with my family. My house is renovated, and you can have your own room. No more sharing the bedroom in the attic at my brother's house." He said, "I have to go. Tourists will be starting to come in for lunch soon."

Soon I found myself at his restaurant, *juxiang lou* (spices gathering restaurant), in Longcun. It was one of only two restaurants providing food on order; the others provided pre-set standard meals to tour groups.

I sat by the cashier's counter. Surrounding the counter displayed a dozen large jars filled with homemade rice liquor, and each was labeled with a different name. I was dressed in San Ge's Miao-style suit with the small embroidery of the restaurant name on the chest. San Ge had two sets custom made, one for himself and the other for his nephew Gui. Gui was the only employee here, and San Ge paid him a monthly salary of 2,400 yuan.[3] Tourists were yet to arrive. I was familiarizing myself with different rice liquor; Gui was sweeping the floor, and San Ge was sitting outside the door, smoking a cigarette (Photo I.1). All were waiting to find out if it would be a good day for business.

We didn't have long to wait. As noon approached, the restaurant became hectic. There were five tables of tourists, four of which were tour groups

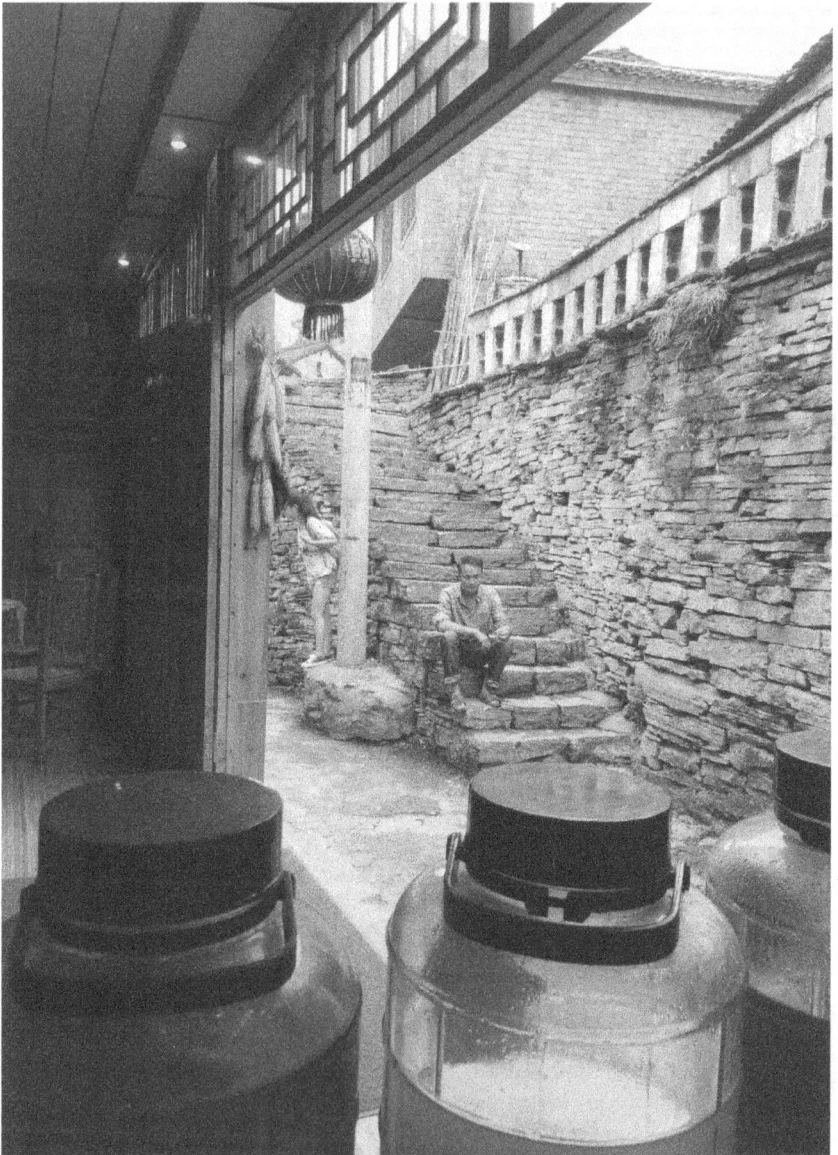

Photo I.1.   Over the rice liquor jars on the counter of *juxiang lou*, I saw San Ge sitting outside, waiting for tourist customers to arrive. (Photo by Xianghong Feng, 2014).

led by tour guides. San Ge was cooking in the kitchen. Gui was running between the kitchen and the dining area, helping with the cooking, serving dishes, and taking orders. I was left in charge of the dining area. One woman splashed oil on her shirt and asked me for help, while two girls wanted me to show them the way to the outhouse. One man was calling me to their table and complaining about the dishes' bland taste, asking if it was because the salt was too expensive here; at the same time a tour guide was asking me to explain to her tour group why the rice cost three yuan (one yuan more than other restaurants). The tourists at one table were signaling me to bring the bill, as soon as those at another table were rushing me to hurry the kitchen.

The last tour group came in after two o'clock. Their guide had originally arranged for them to eat at another restaurant with pre-set standard meals. But they insisted on food on order, so she took them here. After they paid the bill, she asked me in low voice when she could get her money. It took me a while to realize what she meant and to find out what percentage was for the kickback.

"I can give it to you now," I told her.

She said quickly, "Wrap it in paper napkin, so my tour group won't see it."

On her way out, she told Gui that it was not easy to communicate with me, as I didn't seem to know much of the business.

Finally, the tourists at the last of the five tables left, and we were ready for our own lunch. San Ge asked me, "You cook or I cook?"

"I cook, and you take a break," I said.

While I was in the kitchen, another group of tourists arrived and ordered three dishes. I hurried to finish my cooking, and gave the kitchen back to San Ge. After the three dishes were served to the tourists, we started eating at another table. San Ge did not eat rice with the dishes. He grabbed a bottle of beer, saying "[I spent] too long in the kitchen, and breathed in too much cooking smoke. I have no appetite."

I noticed Gui hissing as he chewed. "Is it too spicy?" I asked, "How come you complained earlier that I did not use enough spicy sauce?"

"I burned my tongue when doing the fire swallowing stunt at the show last night," Gui replied.

Gui played drum for the campfire-evening show at the World Phoenix Hotel in Tuo River Town at night. He used to perform for the show in Longcun before it was discontinued. "Didn't you usually play drum?" I wondered.

Gui said, "Right, but last night I had to substitute for someone else. I haven't done the fire swallowing stunt for a while, and was nervous."

After eating and cleaning up, Gui and I drove with San Ge to Tuo River Town. After dropping Gui at the World Phoenix Hotel for his evening job, San Ge and I went to replenish the restaurant's supplies. With almost 2,000 yuan spent at an agricultural wholesale market, San Ge and I loaded up his

MiniPassenger van with 30 frozen chickens, 30 *jin*[4] of smoked pork, boxes of eggs, and bags of cabbage and water spinach. We were at the market the day before, and had left with 30 chickens, 20 *jin* of smoked pork, 40 *jin* of Chinese cabbage, and three carts of cucumbers. Most of those supplies had been for the restaurant operated by San Ge's oldest brother, Jing. Jing ran one of the restaurants providing pre-set standard meals to tour groups, whose family I stayed with during my previous visits in Longcun. That day, Jing's restaurant had 270 guests, many more than San Ge's. "Even though my brother had many more tourists, with only a 3–5 yuan profit margin per tourist he made the same money as me," San Ge explained.

While we were in Tuo River Town, a girl was sent by her tour guide sister to pick up her kickback for the tour group she took to *juxiang lou* that day. "I give 40 percent of the tourists' expenditure back to their tour guides," San Ge told me. "The tour guides try to take their tour groups to those standard-meal restaurants such as my brother Jing's, because they made larger kickbacks there, with 40 yuan out of 50 yuan per tourist." After paying the kickbacks, *juxiang lou* had a gross income of around 1,000 yuan that day. San Ge remembered the best business was on the Labor Day holiday in May. *Juxiang lou* received 40 tables of tourists, with a gross income of 12,000 yuan.

I was excited for San Ge, despite the uncertainty in *juxiang lou*'s future. He seemed one step closer to his dream. I first met him on a summer evening in 2011 while conducting fieldwork in Longcun village in Fenghuang County of Hunan Province in China (Figure I.1). At that time, he was a self-employed rental van driver, who managed to start his first small tourism business by using all of his eight-year savings from wage earned as a migrant worker combined with some borrowed money. Starting as a farmer tied to his fields, to a migrant worker, then a tourist rental van driver, back to being a migrant worker, and now opening his own tourist restaurant, San Ge had struggled, with enduring resilience, to take part in the seemingly prosperous tourism market in Fenghuang, just as many other ordinary peasants in local communities.

Tourism is a major force to bring change worldwide. In China, urbanization through tourism is a common strategy for the state and local authorities (Nyiri 2006). From being a "backward" ethnic region historically to where the KFC fast-food chain opened its first store in 2013, Fenghuang's urbanization through capital-intensive tourism development since 2002 offers an illuminating case of economic and sociocultural processes occurring globally, where the poor population is changing, and the meaning of being poor is changing. "Poverty is not a certain small amount of goods, nor is it just a relation between means and ends; above all it is a relation between people" (Sahlins 1974: 37). When economic growth takes the reins, it "creates joblessness and excluding swathes of populations" (Beck and Beck-Gernscheim

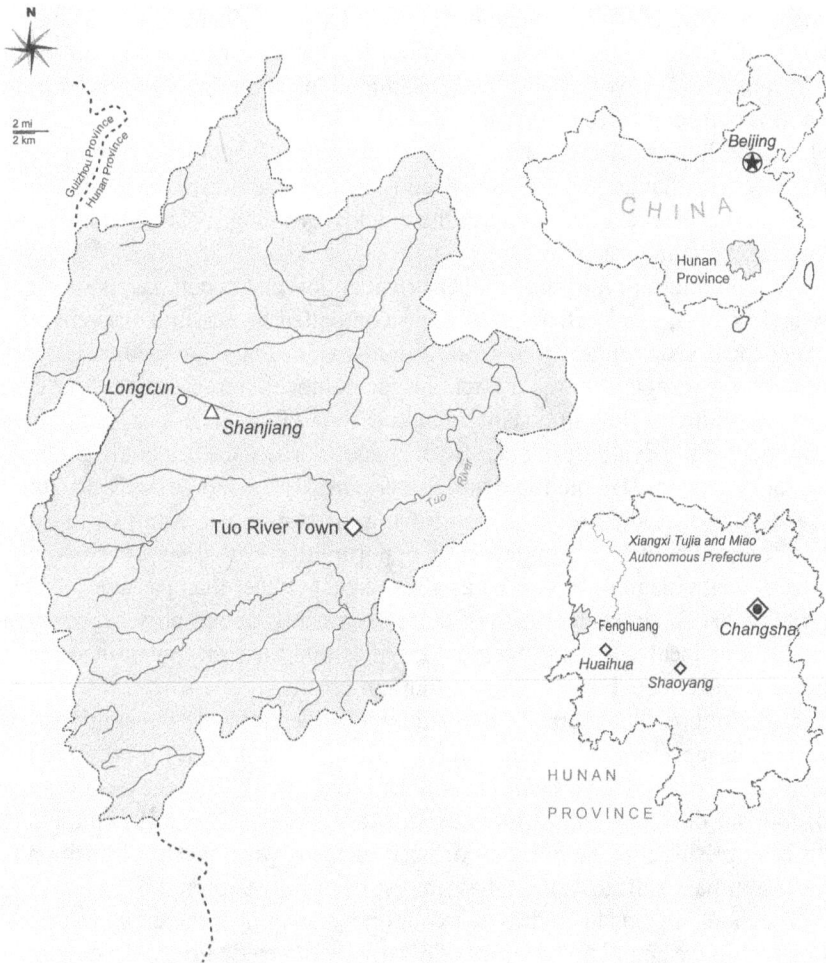

Fenghuang County

Hunan Province, China

Figure I.1. Fenghuang County is located in the Xiangxi Tujia and Miao Ethnic Groups Autonomous Prefecture of Hunan Province, China.

2002: 211). Along with the growth of a flexible deregulated market economy, the poor and excluded, like San Ge, are often left to flounder alone.

In China, with the weakened collectivist structures and a widespread collapse of belief in socialist ethics among individuals who are caught up in the feverish chase after moneymaking, there is an ongoing process of private responsibility, requiring ordinary Chinese to take their life into their own hands and to face the consequences of their decisions on their own (Hansen and Svarverud 2010; Kleinman et al. 2011; Ong and Zhang 2008). The new ways of living reveal dynamic possibilities for a reorganization of social relations, many of which have been examined in scholarly monographs from various perspectives (e.g., Osburg 2013; Rofel 2007; Yan 2009; Zhang 2010). However, thus far few have looked at these changes from the prism of tourism,[5] despite that amid China's hurried remaking and drastic transformation, tourism has played a major role in the quickly evolving social dynamic in its vast rural communities.

As a manifestation of the broader privatization that encompasses much of today's China, rural officials are often compelled to maximize revenue by using local state authority to foster "tourism dynamos" at scenic sites nationwide, packaging tourist attractions for nonlocal entrepreneurs who then extract profits to flow in distant circuits (Zinda 2015). The local authorities contract with private developers to manage tourism sites, including ethnic minority villages (Blumenfield and Silverman 2013; Chio 2014; Nyiri 2006; Oakes 2006). And there has been growing interest in discussing various actors' engagement and negotiation of opportunities in tourism development, where local villagers are viewed as active agents rather than passive subjects (Nitzky 2013). However, much of these discussions do not provide a critical analysis of local villagers' everyday politics and their discontent and resistance to the unequal distribution of tourism profits.

Furthermore, recent contributions to a debate concerning the continuation versus disappearance of the peasantry called for the need for more in-depth analyses of peasants in transition (Van der Ploeg 2008). With the proliferation of tourism development in rural contexts, as Tucker (2010: 928) points out, "it is surprising that the question of what happens when peasant transition is a predominant shift to tourism has not yet been fully explored."

This book intends to contribute to our understanding of peasantry in transition within the complex dynamics of externally directed tourism development and the changing social and economic relations that accompany such development. By focusing on the intersection of tourism, power, and inequality in today's China, this book: (1) examines developmental issues within the specific context of tourism in its interior south, where the majority of local residents are the ethnic Miao;[6] (2) assesses how large-scale and capital-intensive tourism development affected the living conditions of the local residents; (3) reveals how these changes created tension, social and cultural disruptions, shaped the aspiration of the residents, and generated uncertainty; and (4) evaluates the implications of this study for a better understanding of Chinese economic and sociocultural dynamics and how this understanding helps decision makers in China and worldwide to shape more sustainable development policies.

In this book, I conceptualize power in tourism as present everywhere in a system consisting of tourists, local residents, and tourism mediators (mainly referring to the political authorities and private tour developers and operators in this book). I examine the role of tourism developers as an emerging dominant power in shaping local tourism and society, and pay special attention to the intricacies of host communities, to show how they might be internally divided with the intrusion of external economic forces. I look into the daily conflicts over access to profitable tourism resources between the political and economic elites and the local peasants. I investigate on one hand how the private tourism developers, connected with local political authorities, were reaping most of the material rewards through the control of the temporal-spatial distribution of tourists, and on the other hand how local authorities' personal wealth was strengthened through their facilitation of the tourism developers' pursuit of profit. I explore how, during this process, the powerful enforced their hegemony, despite local resistance, resulting in an intensification of the marginality of the powerless.

As a fundamental social force that assembles a broad array of social, political, economic, cultural, and material processes, tourism serves as an analytic lens through which new insights into these processes, and thus our theorizations about them, are gained (Minca and Oakes 2011: 1). As stated by Silver (1999: 504), "what is interesting about tourism is not tourism per se, but rather what tourism represents in the context of our modern world . . . this peculiar activity can teach us a great deal about the wider phenomena of domination, subordination, and resistance."

Ethnographically grounded, this book strives to stimulate a greater connectivity between general social theories and tourism analysis, acknowledging that tourism study is an innately interdisciplinary field. It relies heavily on anthropological concepts and methods, while drawing inspiration from other social theories in sociology, geography, and political science (e.g., the McDonaldization thesis, spatial theory, and everyday resistance), to understand the role of tourism in human lives. It emphasizes contextualizing and advancing theoretical understandings on scale, space, gender, and their dialectical relationships with power through an analysis of tourism in this given locality. These theoretical underpinnings are subtly embedded to place the collective experiences of those depicted in this book in perspective and give them meaning.

The local situation depicted here is far from unique, as it concerns the relationship among agriculture, outmigration, and tourism-related entrepreneurial endeavor, and unravels the process of urbanization and individualization expedited by tourism growth. It has broader implications, and is of value as a contribution to sustainable development in tourism and in general.

DOING ANTHROPOLOGY IN FENGHUANG

During the Chinese New Year in 2002, I left my hometown in Shaoyang for Fenghuang to gather data on local Miao women's traditional handicrafts for my master's thesis in folklore. This first visit immediately preceded Fenghuang's large-scale tourism development. I later developed an interest in Fenghuang's tourism as my anthropological dissertation project and conducted ethnography in 2005 and 2006. My dissertation (Feng 2008a) evaluated the preliminary economic and sociocultural impacts of this capital-intensive tourism development jointly promoted by the county government and an outside for-profit cooperation. It found that only a minority of the local people shared a small proportion of the economic benefits, and the majority of them were paying the costs, such as environmental degradation and social conflicts. I concluded that input from the more developed region, instead of helping a developing small-scale society, actually might harm it, because reliance on outside capital led to dependence, wealth extraction, structural inequalities, and resentment among the dependent people (See Feng 2008b for a condensed analysis).

This book is, however, based on the data collected from my later fieldwork in Fenghuang between 2010 and 2014. Participant observation and interviews were mainly carried out in Fenghuang's urban and tourist center of Tuo River Town, and three of its rural communities (Longcun, Wucun, and Huangcun) with popular tourist attractions. Each of them has its own unique situation, offering insights into the dynamic interplay of different stakeholders in the local touristic system. The interviewees, representing various aspects of local life and tourism, were selected either purposefully because of their roles (e.g., local political authorities, private developers, itinerant merchants, and local peasants), or incrementally through personal contacts. Some of them have been my long-term informants since my first visit in 2002, and were interviewed more than once and on different occasions.

Daily casual conversation, on-site observation, and locally accessed archives (including internal reports and other unpublished materials from the county government, village cadres, and tour operators) provide additional valuable information. Phenomena were examined from multiple perspectives and using varied sources of information. Special efforts were made to cross-check references from different stakeholders with conflicting interests. Interview data were either recorded through written notes or audio recording if permitted. The interviews were conducted often in Mandarin and the Hunan dialect in which I am fluent, and sometimes in the Miao language with assistance from a locally recruited interpreter, who was a student at a local vocational school and then interned with a tour agency in Tuo River Town.

Drawing upon ethnography spanning over a decade in Fenghuang, I explore in this book the complexities of life in contemporary rural ethnic China as China undergoes one of the fastest social transformations in history. This is also an intimate personal story of my life's journey in Fenghuang, as an anthropologist, a camerawoman, and a restaurant helper, or as an adopted daughter/sister to my host families, and a friend to many of my informants. Living there and traveling on foot and by mini-bus, motorcycle, and motor rickshaw, my daily encounters included interactions with tourist and the toured, local and non-local, Miao and Han, men and women, young and old, from local officials to peasants, from private tourism developers and operators to tour guides, from itinerant merchants to repatriated migrant workers. Multiple narrative threads are woven into illuminating depictions of the place and its people confronted with a rapid pace of change. The stories are structured around those I encountered, and reveal how ordinary people responded to rapidly shifting opportunities and constraints brought by tourism.

This ethnographic monograph particularly focuses on the changing conditions of local life in the context of a new tourism economy, and presents a collection of personal realities of those whose lives have been affected by tourism. We have met San Ge, Jing, and Gui at the beginning of this chapter. San Ge has two older brothers (Jing and Ting), and one older sister (Gui's mother). They each established their own households. While Ting moved to the market town of Shanjiang, San Ge, Jing, and their sister's families remained in Longcun. Other main characters in Longcun who appear later include Qiu (the Village Communist Party Secretary), Zhi (the former Village Communist Party Secretary), and Mi (the long-term Village Treasurer). Lu was the developer of Longcun as a Miao village tour site, and he is from outside of Fenghuang.

Ma was the developer of Wucun, another Miao village tour site, and he is a Wucun resident. While Lu represents the outside private tourism developers, Ma represents the local developers. Ma's management team includes Pang Ge and Ping, both of whom are from Wucun, and Ping was the former Village Communist Party Secretary. Ma had Ping supervise daily tour operations in the village, and placed Pang Ge in charge of the tourist reception office in Tuo River Town. Hai, another Wucun villager, was once a member of Ma's team, directing Ma's performance troupe that held daily shows to entertain village visitors. Over the years, Hai and his wife along with their adult children gradually moved from Wucun to Tuo River Town to start their own small tourism businesses, including their two younger daughters Xia and Yun. Shun was Xia's husband, who is from Tuo River Town.

Jun was the co-owner of a major local travel agency. Shi was one of his senior staff in its office located in downtown Tuo River Town. Hua was one of the tour guides associated with Jun's travel agency, who later became my Miao language interpreter and research assistance for two summers. Liang was a camerawoman, and lived with her husband Hong and their toddler son in a rented room in Tuo River Town, away from their village home. Ru and Teng were both officials at the county Ethnic Affairs Bureau. Teng joined the bureau right before Ru's retirement. Yao was with the county Statistics Bureau. Most of these individuals and their families are introduced for the first time in Chapter 1, with many reappearing throughout the volume as key informants.

Using contextual knowledge to interpret personal experiences, the individuals portrayed in this book and their stories are not presented as stand-alone cases. Their lived experiences reflect larger outside forces impinging upon them and their reactions to these forces, and shed light on the linkage between the changes at the macro level of local, national, and global development policies and the economic dilemma faced by them at the micro level. While numbers obtained through locally accessed archives show the overall trends and patterns, the lived experiences from all walks of life in Fenghuang are especially valuable to understand peasantry, poverty, and social and gender inequality in contemporary China.

SCOPE OF THE BOOK

What this book reveals concerns tourism, and more importantly, the social world and how it is changing. Over the last twelve years in Fenghuang, the introduction of large-scale and capital-intensive tourism development has involved the cultural reconstructions of space, ethnicity, gender, and morality within changing configurations of power relations. In the transition from subsistence agriculture to wage labor and small-scale entrepreneurship, local peasants have reformulated their preexisting meanings, values, and practices through their everyday enthusiasm, acquiescence, or resistance to a tourism-dominated market economy.

Organized into three parts, the following chapters proceed chronologically beginning with the local residents' encounter with tourism, spreading from the capital town to the countryside. Meanwhile, each chapter (except Chapter 1) utilizes a specific theoretical framework to examine relevant ethnographic data that are organized around a particular thematic anchor. From different aspects, they together contribute to the investigation of how the private tourism developers, connected with local political authorities, were reaping most of the material rewards and how, during this process, the powerful enforced their hegemony, despite local resistance, resulting in an intensification of the marginality of the powerless who are local residents, especially the Miao peasants.

Part 1 (Chapters 1 and 2) introduces how Fenghuang County's large-scale and capital-intensive tourism development, centered on Tuo River Town, took the "Great Leap Forward," and how this development has spread to the county's rural communities with the advent of a nationwide promotion of *"xiangcun you"* ("countryside tours"). Chapter 1 takes readers to Fenghuang. It contextualizes Fenghuang and the Miao geographically and historically; it discusses how Fenghuang became a tourist destination, and introduces tourists, tour agencies, developers, and guides, traders, and local officials, many of whose stories are elaborated in the following chapters.

Chapter 2 critically applies a sociological perspective known as the "McDonaldization thesis," and examines three cases of village tours (the "overpackaged," the "abandoned," and the most "popular"). It documents the experiences of the toured and tourists, and their response to global forces of commodification, dehumanization, and alienation. It examines the role of tour operators (both local and non-local) as an emerging dominant power in shaping local tourism and society, through which both the toured and tourists are being objectified and consumed for profit at the hands of the global market. The local situation reveals the dynamics of the global process of "scaling up." It shows how during this process the private tourism operators and developers, connected with the local political authorities, were reaping most of the material rewards through their control of the temporal-spatial distribution of tourists.

Part II (Chapters 3, 4, and 5) follows the penetration of tourism into rural Miao communities, and examines on one hand how these communities are incorporated within the local tourism system, and on the other hand how they are largely excluded from the actual tourism sites and, therefore, from profits and other benefits. Chapter 3 expands the discussion in Chapter 2 on the temporal-spatial control of tourists by the tourism mediators, to see how power and social transformation are embedded in tourism sites. It links the production and reproduction of power with tourism sites, and examines the material spatial practices in the progress of Fenghuang's urbanization through tourism, during which the local public officials and outside private developers acquired control of spatial, therefore, socioeconomic, order. Placing spatial practices center stage in touristic power relations, it looks into the daily conflicts over the access to profitable tourism resources between the political and economic elites and the local residents. Relevant illustrations are drawn from the experiences of street vendors in the *Gu Cheng* (the historic part of Tuo River Town), as well as the construction of the performance stage and vending stalls in Longcun village.

Chapter 4 presents cases of individual Miao women and men at the local tourism marketplace and in their village homes, to reveal the changes in gender division of labor resulting from the shift from agriculture to wage labor, and then to the growing tourism economy. Their stories demonstrate an increased flexibility of women's roles but suggest that this increased flexibility might best be interpreted as a result of Miao women, who occupy the low end of every form of power in the local and national social structure, being pushed around by other more powerful internal and external agents. This increased flexibility did not necessarily empower them, but might instead maintain their subordination to men under changing socioeconomic circumstances. The traditional gender ideology of men's superiority was therefore not challenged but reinforced with the penetration of market economy.

Chapter 5 explores the dilemma faced by many peasants in today's China: they have to struggle to live off the land and seek security beyond agriculture, while the land is the only real source of security. Through cases of three Miao men (representing both ordinary villagers and village cadres) and their endeavors as small-scale tourism entrepreneurs, it investigates the relationship between agriculture, migration, and small-scale entrepreneurs among the local Miao peasants, and documents the experiences of those who were involved in tourism-related entrepreneurial endeavors. It assesses the choices and constraints of these local lower-class Miao in their attempt to shift the primary source of family income from labor to small-scale capital. The emphasis is on the Miao peasants as actors who tried to transcend their different, yet similar, plights. Bringing together the discussion in the previous chapters, it suggests that their current living condition is better described as the "poverty of resources"—the erosion of resources and social protection through an exclusive economic development model that features the privatization of previously communal resources—and that the key to bringing about equitable growth depends on a more moral, rather than a solely market, allocation of resources.

Part III (Chapters 6 and 7) shifts the focus to tourism-induced internal dynamics among the local Miao peasants in one village of Longcun. Chapter 6 focuses on the everyday resistance, where the villagers were defending their interests in tourism benefits. It spotlights the informal and often individual resistance acts, and presents a series of entangled incidents surrounding the vending stalls, tourist restroom, roadblocks, and performance show. It reveals how particular resistance incidents evolved among these villagers, and pays attention to their internal politics and the cultural underpinnings of their resistance acts. Responding to various situations, these Miao peasants made rational choices among possible trajectories of action. If the resistance was direct, it was not open; if it was open, it was not direct. As actors with little social power, they worked the existing system to their advantage: they cleverly utilized policy gaps to persist in their resistance; and they strategically used tourists as the scapegoat to indirectly confront the dominant. Despite the marginality and temporal nature of their resistance, it did affect the various forms of exploitation that they confronted, and therefore was significant in the sense that it narrowed the options available for the dominant.

Chapter 7 revisits the traditional peasant society worldview of the "limited good" that sees social division of wealth holistically as a zero sum game, and its associated concept of "competing to remain equal" through the case of Longcun. It discusses the cooperation and competition among the village family restaurants, and explores the rationality behind economic behaviors of these peasant entrepreneurs. It argues that peasant entrepreneurship, as observed in Longcun, was influenced by an "image of limited good" and the prominence of "jealousy." In the context of tourism development, the local peasants might have adopted elements of official "development discourse," and desired to achieve material affluence. But the current development model, which is derived from the globalized society worldview of the "unlimited good" and "competing to get ahead," advocates individual rather than collective prosperity, and deepens social inequality. It contradicts the traditional peasant worldview of the "limited good" and "competing to remain equal." Influenced by the "limited good" viewpoint, social pressure based on egalitarian morality (e.g., the prominence of jealousy) in peasant societies like Longcun has worked as an effective mechanism to strive for not only prosperity, but also equality.

I left Longcun in mid-August that summer. San Ge called me three days after I left. It was in the evening, and he was again in Tuo River Town to replenish food supplies. He was done shopping, and was waiting for the vendors to deliver the goods to his van parked at the market entrance. There were many visitors in the village that day, making it a good business day. With over 100 guests, his restaurant had almost 2,000 yuan of gross income after paying the kickbacks. He was so busy with cooking that some tourists ordered dishes first and then went to take a walk around for an hour before coming back. He told me that he just hired a twenty-year-old girl from another Miao village with an offer of a monthly salary of 1,500 yuan, and she would start working the next day. A year later though, I heard that the restaurant no longer had much business, and he had left the restaurant to Gui. San Ge was commuting to Tuo River Town daily to do carpenter work for shops.

NOTES

1. Except for the Deputy Head and the Communist Party Secretary of the county government of Fenghuang, all people's names are pseudonyms in this book. San Ge means "the third older brother" in Chinese, which is how I addressed him in person. It is used here as a pseudonym as he wanted.

2. The village, which I call Longcun, is not its real name. The same is with other village names in this book.

3. The exchange rate of US dollars to Chinese yuan was about 1:6 in 2014.

4. One *jin* equal to 0.5 kilogram.

5. Among the monographs on tourism in China, Oakes' (1998) examines from a cultural geographical perspective the theoretical conceptions of place, modernity, and authenticity, and elaborates on the power relations between the ethnic periphery of Guizhou, populated by the Miao, and the centralized power of the Chinese state. Also from a geographical perspective, Su and Teo (2008) conceptualize power relations in tourism space, and unravel the politics surrounding China's hegemonic project of heritage tourism development in the case of Lijiang in Yunnan. Through textual analysis and limited ethnography in three "scenic spots" in Sichuan, Nyiri (2006) draws a general picture of the state of Chinese tourism at home and abroad, and explores how tourist sites are constructed in China by the state. Surrounding the issue of authentication in ethnic tourism, Xie (2011) traces the changes of tourist folk villages, peoples, and tourism landscapes among the ethnic minority Li in Hainan of China. Shepherd (2013) uses the example of China's Mount Wutai, a UNESCO World Heritage site and a national park, to examine the political, economic, and social foundations of the world heritage process. Most recently, drawing on ethnography in a Zhuang village in Guangxi and a Miao village in Guizhou, Chio (2014) investigates two social processes (mobility and visuality) that shape the "work" of tourism in rural ethnic China.

6. Miao is one of China's fifty-five officially designated ethnic minorities, referring to the various groups of people in Southwest China who call themselves Hmong, Mong, Hmu, or Xioob. Their languages are closely related, but their dialects, customs, myths, and legends are diverse from place to place. The related people in Vietnam, Laos, and Thailand are usually designated as Hmong or Meo (for an elaborated discussion on defining the Miao, see Diamond 1995). In Fenghuang County, slightly more than half of its total population (58 percent of 423,032) is Miao, who largely concentrate in its rural areas (Fenghuang County Statistics Bureau 2013). See Rack (2005: 18–19, 21–44) for a full discussion on how the category "Miao," particularly the Miao of West Hunan, has been constructed historically and its relevance today.

Part I

TOURISM'S "GREAT LEAP FORWARD"

Chapter One

The Place

The Mountain, The River, and The People

If one looked at an old map from a hundred years ago, they could surely find in an extremely remote corner among northern Guizhou, eastern Sichuan, and western Hunan, a little dot called *Zhen Gan* (Shen 1986). *Zhen Gan* is today's Tuo River Town. Sitting by the midstream of Tuo River, a tributary of the Yangtze, it is the county capital of Fenghuang. *Fenghuang* means phoenix, the mythological and auspicious bird in China. About fifty miles west of Tuo River Town, there is a bird-shaped mountain called "Phoenix Mountain." Fenghuang County was thus given the name (Fenghuang County Ethnic Groups Gazetteer Writing Committee 1997).

Even though few ethnographic data (except Rack 2005[1]) were drawn from West Hunan in recent anthropological scholarship on ethnicity and China's southwest frontier zones,[2] it is a well-known place in Chinese modern literature, due to the popularity of the writings of the distinguished author Shen Congwen, himself partly of Miao descent from Fenghuang. Shen (1982, 1986) portrayed his home area in West Hunan for its particularity in habits, manners, folklore, and beliefs. In writing of his hometown in "Fenghuang," Shen (cf. Oakes 1995a: 100) notes: "This is Miao territory in the borderland, and these semi-primitive people's belief in spirits exerts a tremendous influence on everyone. There are spirits everywhere, in trees, caves, and cliffs. . . . Because people love each other and moral concepts are extremely strong, legends arise about love between mortals and spirits or monsters, and women find an outlet here for their sexual repression." To the outsiders, West Hunan has long been imagined as an isolated and somewhat mythical place, equipped with images of exotic, sexual, and "dangerous" Miao (Diamond 1988), as depicted in Shen's work. Shen's work contributes to tourists' "geographical imaginaries" (Salazar and Graburn 2016: 4) and "sexual imaginaries" (Salazar and Graburn 2016: 9) of this place and its ethnic minority people.

According to Oakes' (1995a) analysis, Shen's work presents West Hunan as a beautiful woman, and valorizes the primitive by celebrating woman as nature, country, and purity. Echoing Kinkley (1987), Oakes (1995a: 99) points out that in Shen's imagination the Miao way is that of the early Chinese, and "the culture of the frontier, found especially in Miao women, represented a vitality which once belonged to the Han." He (1995a) further argues that Shen portrays the frontier folk as the bearers of China's cultural residue as the Miao still practiced certain ancient Han customs long forgotten by the Han themselves. Influenced by Shen's work, many desired to visit Fenghuang as a romantic country with universal qualities of an idyllic life in relative isolation, to satisfy their urban-induced cultural nostalgia as an antidote to modernization.

Tuo River Town was named after the river running through the town. The *Gu Cheng* (literally means "ancient town"), referring to the historic part of Tuo River Town, was an irresistible attraction for the urban tourists for a look back at a time and place gone by. Historic city walls and gates surrounding the *Gu Cheng*, narrow cobbled streets dotted with shops displaying local colors and flavors, the clear river ever flowing slowly downstream, weathered wood-stilted houses hanging over both sides of the river, jumping rocks (*tiao yan*) connecting one riverside to the other, small boats traveling up and down the stream, and the day starting with the rhythm of local women pounding clothes with wood beaters by the riverside doing laundry in the morning mist (Photo1.1). Resembling the image of the town of Chadong in Shen's (1934) most idyllic and popular work *Bian Cheng* (Border Town), these were once the natural landscapes of West Hunan.

With modernization as a state ideology in China, the entire gradual penetration of the outside world into the ethnic inland was inevitable (Oakes 1998; Schein 1997; Swain 1990), and Fenghuang was no exception. The rising ethnic tourism growth in Fenghuang further intensified the existing commercial penetration well beyond its capital town into the remote rural communities of the Miao. As a spiritual journey for the metropolitan Chinese, Fenghuang has become a popular destination for painters, musicians, photographers, and ethnographers who come to seek inspiration for their urban pursuits or in search of a firsthand experience of rural otherness. Fenghuang, as the exotic, the traditional, and the "pre-modern," was an ideal place to promote culture tourism, with the picturesque *Gu Cheng* and the "primitive" Miao villages as its most profitable tourism space. This chapter contextualizes Fenghuang and its large-scale tourism development, and it introduces the tourists, tour agencies, developers, guides, traders, and local officials, many of whose stories are elaborated in the following chapters.

Photo 1.1.   Overlook of the *Gu Cheng*, the historic part of Tuo River Town. (Photo by Xianghong Feng, 2010).

FENGHUANG AS PERIPHERY

Fenghuang County is located in the southwest corner of the Xiangxi Tujia and Miao Autonomous Prefecture in Hunan Province, where it borders neighboring Guizhou Province to the west. Along with Guizhou, Fenghuang occupies the heavily eroded limestone highlands of the eastern Yungui Plateau, the hinterland of southwest China where the Chinese Miao ethnic groups are largely concentrated. Fenghuang and the prefecture share more in common economically and culturally with their neighboring regions in Guizhou than Hunan, due to geography, physiography, and ethnic composition.

China's sub-division into provinces took shape during the Yuan Dynasty, and its essential principle of basing provinces on administrative rather than economic considerations has remained in force ever since. For centuries, there existed a separation between economic and administrative boundaries. During this time, the central provinces remained relatively stable and well defined, albeit with frequent minor rearrangements and border corrections (Fenghuang County Ethnic Groups Gazetteer Writing Committee 1997). In contrast, economic regions were defined in different ways for different purposes. For late Imperial China

(China's Ming and Qing Dynasties, roughly between 1400–1900), Skinner (1964, 1977, 1985) combined natural features and market structures to define nine macro-economic regions that generally cut across provincial borders.

According to Skinner (1985), China had historically been characterized by a hierarchy of local and regional histories whose scope was grounded in the spatial patterning of human interaction. Skinner grouped these local and regional histories into nine physiographic macro-regions, each characterized by a core area of agricultural accumulation and urban networks. Skinner (1994) had since used more recent and further disaggregated data to argue that these macro-regions were still relevant to the modern Chinese economy (Hendrischke 1999). Oakes (1998: 87–89) employed Skinner's macro-regional system to analyze Guizhou's position as China's physiographic periphery.

Despite the limitation of this macro-region model argued by Cartier (2002), it is appropriate to apply it here for the purpose of understanding Fenghuang's peripheral position (similar to Guizhou) in a historic context. Looking at the administrative region defined as Fenghuang on the map reveals its peripheral nature in physiographic terms: it is entirely a hinterland region, located in an ethnic minority autonomous prefecture, close to the borders of no less than four of Skinner's nine macro-regions (Figure 1.1).

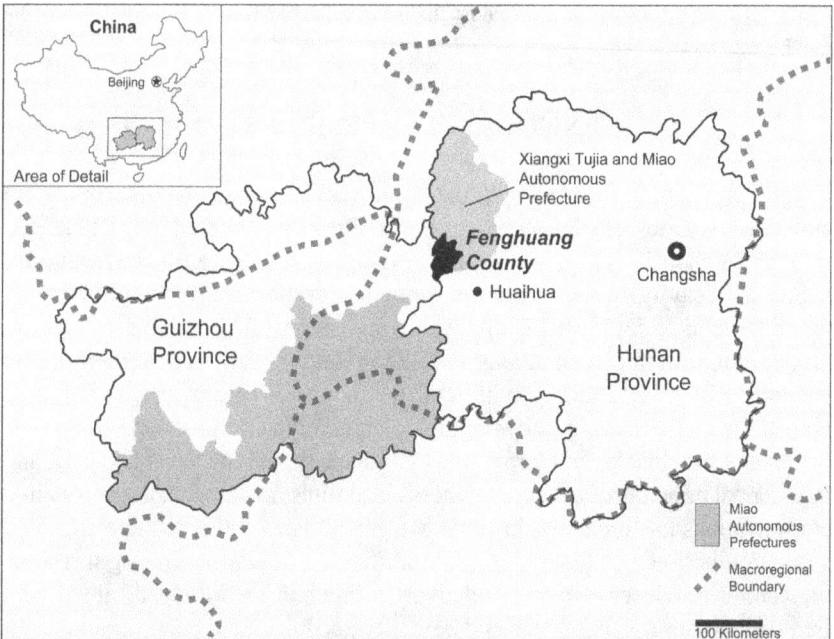

Figure 1.1.   Fenghuang County's location relative to the peripheries of Southwest China's physiographic macro-regions. (Data adapted from Skinner 1985: 273).

Just as how Oakes describes Guizhou (1999:31), the mountainous setting of Fenghuang is "responsible for a long history of not only relative isolation from the dominant societies and cultures of traditional China but of endemic rural poverty and a stigmatized 'frontier' cultural identity as well." Scholars (e.g., Goodman 1983; Oakes 1998; Spencer 1940) have characterized Guizhou as an "internal colony," because of unequal resource extractions first by the Chinese state, and more recently, by national level Chinese capital. The "internal colony" concept is applicable to Fenghuang as well. The history of this region has been a regional history of continuing cycles of conquest, revolt, and trade, a history of conflict and cooperation with the central government, especially during the Ming and Qing Dynasties. The long history of imperial, republican, and communist efforts in Fenghuang have been to tap its resources and integrate the region politically, economically, and culturally into the state (Fenghuang County Ethnic Groups Gazetteer Writing Committee 1997; Fenghuang Ting Local History 2003; Shi 2002).

Despite alternating attempts at their destruction or assimilation by the state, the segmentary tribal organization of the Miao had allowed them to preserve their language and culture for over two thousand years (Lemoine 1978, cf. Tapp 2003:3). Tapp (2003:3) states, "The preservation by the Miao of their ethnic identity despite their being split into many small groups surrounded by different alien peoples and scattered over a vast geographical area is an outstanding record paralleling in some ways that of the Jews but more remarkable because they lacked the unifying forces of literacy and a doctrinal religion and because the cultural features they preserved seem to be so numerous." It is the well-preserved Miao cultures, along with the picturesque mountains, river, and historic town that endow Fenghuang with its unique charisma.

FENGHUANG AS TOURISM DESTINATION

Fenghuang was once one of the "state-level poverty-stricken counties." In 2000, the gross domestic product (GDP) per capita in Fenghuang was only 29 percent of the national average (Chinese Academy of City Planning and People's Government of Fenghuang County 2005). As with China's other peripheries, poverty in Fenghuang remains conceptualized in official discourse as a problem of an incomplete modernization, and tourism is seen as an ideal catalyst to stimulate its economic growth through external investment and market integration. Since 1978 (especially after Deng Xiaoping's South China Tour in 1992), China's post-socialist turn toward a market economy has coincided with a parallel shift to neoliberalism in the global capital world.

The use of the term "neoliberalism" here is not an uncritical application but an appropriate model for China's recent development. As Tang (2014: 44) points out, China's post-reform market development is indeed a kind of neoliberalism. Critical of Harvey's application of neoliberalism to China in his coinage of "neoliberalism with Chinese characteristics," Tang (2014: 53) cites Chu and So's (2010: 49) discussion regarding the major difference in analyzing neoliberalism between China and the United States, and argues that in China it is the communist party state, rather than the capitalist class (like in the West), that has been the dominant agent of neoliberalism. China's neo-liberal approach has involved the fusion of local and outside capital interests, and has transformed local government through the drive to attract outside capital (Walker 2008). This process has provided the backdrop for, and in a large sense shaped, Fenghuang's recent development. During the past decade, large-scale and capital-intensive tourism development has been aggressively promoted by the local government as a solution to the impoverished conditions in Fenghuang.

For Sale

Since the 1990s, the county government has tried to increase local revenue through tourism. However, as many of the visitors were from upper-level government offices and state-owned enterprises, the local officials had to host them with "wined and dined" tourism receptions at the county government's expense. According to the county government, in 2001 there were 130,000 tourists during the May "Golden Week" (a weeklong national holiday celebrating Labor Day), which brought about 30,000 yuan[3] in revenue, but with a reception cost of 400,000 yuan. To be relieved from the burden of reception tourism, the county government decided to put the previously state-managed local tourism resources into the invisible hands of the market, and actively looked for outside investors with both deep pocket and relevant expertise.

Some scholars believe that the introduction of market mechanisms to privatize heritage resources as tourism attractions goes against the public benefit mission of heritage management, while others argue that the government lacks the funds and expertise to effectively manage tourism attractions, especially regarding profit-making (Tang 2013, also see EDRCCASS 1999; Li 2002; and Yang 2004). Despite this controversy, in November 2001 the Fenghuang County government leased the exclusive rights to manage eight major tourist attractions for 50 years to the Yellow Dragon Cave Corporation (YDCC), headquartered in the provincial capital of Changsha City, at the price of 830 million yuan. Five of the eight attractions are located in the *Gu Cheng* within Tuo River Town, including River Boat Sightseeing, the historic homes of Shen Congwen, General Xiong Xiling, and the Yang family, and the historic East Gate. Among the other three attractions, Qiliang Cave is located just north of Tuo River Town, and the Southern China Great Wall and Huangcun are 10 miles and 15 miles to the west of Tuo River Town, respectively.

The YDCC set up a subsidiary in Tuo River Town, the Phoenix Ancient Town Tourism Co. Ltd (PATT), to run the business. The county government commissioned experts from Beijing to make an ambitious blueprint for the county's eighteen-year (2002–2020) urbanization-through-tourism plan (CACP and PGFC 2005; TPCCACP and TBFC 2004). In this elite-imposed process, deals were sealed without input from the local residents. Local residents were left with little choice but to embrace this large-scale and capital-intensive tourism development. One Huangcun villager said, "I had no idea that the government sold our village rampart to the tourism company until one day I was shocked to see the tourists touring on top of the rampart outside my door."

Following the deal with the YDCC, the county government gradually leased additional tourism attractions and land to more private developers. Since 2002, Fenghuang's tourism has gone through two main stages: (1) first concentrated on Tuo River Town (especially the *Gu Cheng* area), monopolized by the YDCC; and then (2) spread into rural Fenghuang through the promotion of countryside tours (mainly Miao village tours featuring ethnic culture and rural scenery), privatized to other tourism developers. Fenghuang's tourism benefited from its geographic proximity (about four-hours' drive) to the Wulingyuan Scenic and Historic Interest Area, a UNESCO World Heritage Site and Hunan's foremost tourist site that has attracted a steady flow of tourists since 1992.

The Tourists

Tourists in Fenghuang are predominantly domestic with a limited budget, and include college students, young professionals, middle and lower-middle income families. Many choose to participate in packaged tours, especially when it comes to the countryside tours, due to the lack of public transportation. Those that opt for the packaged tours tend to be more constrained by the trip, in contrast to those affluent and independent tourists elsewhere who are more likely to operate on their own terms when traveling. Thus, it is appropriate to consider the tourists in Fenghuang as vulnerable targets rather than powerful agents. The local residents were well aware of their budget orientation and bargaining habits. One of my informants once commented, "those tourists in tour groups might have a budget of a few hundred yuan to spend after paying for their tour packages; if they had ten or twenty thousand yuan to spend, why would they join tour groups?"

Due to the YDCC's successful marketing, the number of tourists in Fenghuang increased nearly fourteen times from 567,000 in 2001 to 8,424,200 in 2013 (Fenghuang County Statistics Bureau 2001, 2013). The local infrastructure was unable to cope with the intensity of tourist visits, especially during peak season. Besides pollution, it resulted in water and power shortages due to excess demand, especially the latter. In Tuo River Town, particularly the *Gu Cheng*, the tourist guesthouses mushroomed, so did the air conditioners installed in each room. The local hydropower electric station was unable to meet the rapidly increasing demand, causing frequent power outages in the *Gu Cheng*. In 2005 while I stayed in a riverside guesthouse for two nights, the lightbulbs dimmed constantly.

According to an official at the county Electric Power Bureau whom I interviewed in 2010, while the overall electricity usage in the county doubled, it increased nearly four times in the *Gu Gheng*. A week before my interview with her, a riverside guesthouse had a fire during one of the power outages. A candle lit the curtain in one room, and the fire spread quickly to the entire wooden top floor that was illegally added by the guesthouse operator. "Our maintenance personnel struggled to make emergency repairs day and night," she said. "Last month, the provincial power grid officials came to visit, and committed to help us with both transformers and the installation of underground cables. But once the construction is underway, it will definitely have negative impacts on tourism."

Nevertheless, the electricity shortage in the *Gu Cheng* remained part of daily life in 2011. I was staying at the County Government Hotel for a few days before traveling to the village. The power went off one morning. It came back shortly, but there was still no power for the air-conditioning. It was a power outage that affected the entire *Gu Cheng*. The hotel had its own generator for emergency basic supply, but many shops were closed. The ones that remained open either used candles, or had small generators running outside their doors. The power went on and off three times that night, one of which caught me in the middle of a shower. Liang, one of my informants who worked as a camerawoman to do tourist photography, called and offered to come over with her husband Hong to pick me up to stay at their place for the night. They had moved from a village in Duli Township and rented one room not far outside the *Gu Cheng* where they still had power.

Tour Agency, Developers, and Guides

As a camerawoman, Liang made money by renting out ethnic costumes to tourists and taking their pictures. When the opportunities arose, she also showed her tourist customers around as their tour guide. Literally, anyone in

Fenghuang could make money as a tour guide, affiliated either with travel agencies or tourism companies, or working independently. Since Fenghuang's tourism boom in 2002, tour agencies, developers, and guides rapidly emerged and gradually saturated the local tourism market, among whom were Jun, Ma, Pang Ge, and those associated with them as depicted below.

1) *Jun's Business*

Ru arrived around noon. The deputy director of the county Ethnic Affairs Bureau, Ru had been my primary liaison in Fenghuang since 2005. It was a Saturday in June 2011. She had just come from an emergency government meeting that morning, as the developer who leased the Tian Shrine in the *Gu Cheng* had conducted unauthorized renovation and demolished its seven cultural relics. "We have been so much busier since the recent tourism boom," Ru complained.

Ru took me to meet Jun, the deputy general manager of a local travel agency, the Xiangxi Fenghuang Holiday Travel Agency. Ru had come to know Jun over the years, as she was in charge of making travel arrangements for the bureau staff's business trips. Jun's travel agency was in a two-story unit, located by the Rainbow Bridge that crossed the Tuo River at the center of the *Gu Cheng*. Jun's wife was the office manager. While Jun was often out of town on business trips, she was responsible for its daily operation, receiving customers, selling tour packages, organizing tour groups, assigning tour guides, and arranging transportations. Jun welcomed tour guides to be affiliated with his travel agency, as long as they requested little or no base salary. Jun paid each tour guide 50 yuan for a one-day group tour. It was an open secret that tour guides relied on the kickbacks to survive, and the amount of the kickbacks depended on the tour guides' ability to make tour groups consume at the businesses (shops, restaurant, guesthouses, etc.) with which they had agreements.

Over the years, Jun's travel agency became one of my hangouts in town for those summer evenings in between my village trips. There I met Shi, a senior office staff at Jun's travel agency. In his late twenties, he was from Huayuan, another county in the Xiangxi Prefecture. He recalled how clear the river was when he first traveled here with friends in 1998. "It was not like nowadays. Too many tourists and too noisy now," Shi said. "Back then the monthly rent for a riverside shop front was only 200 yuan, and still no one wanted it. Now it is tens of thousands!"

Summer evenings were the busiest time at Jun's travel agency. Steady streams of tourists left the office staff with no break. Jun provided them takeout dinner delivered to the office by a nearby restaurant. One evening during dinner, a middle-aged woman came in to ask for help selling two campfire-evening show tickets to tourists. As one of the local *ye daoyou* (wild tour guides) who had neither license nor affiliation with any travel agencies or tourism companies, she sought out tourists to purchase her services as well as show tickets or tour site entrance tickets that she then sold for commission. Right after she left, a middle-aged man popped in. He was a local salesman for the tour site Heaven Dragon Valley, and his job was to solicit tourists in the streets or at guesthouses to buy its entrance tickets. With a base salary of 600 yuan per month paid by the tour site developer, he made an additional 5 yuan for every ticket sold.

That evening, Jun's travel agency sold six campfire-evening show tickets. The show was held every night at the World Phoenix Hotel (the only four-star hotel, and the best one in town) and was the one for which Gui worked as mentioned previously. Jun's wife assigned Hua with the task of taking four tourists to the show. Hua was one of Jun's tour guides, a local Miao girl from Sangongqiao Township. I later recruited Hua as my Miao language interpreter for two consecutive summers. The official ticket price, as printed on each ticket, was 228 yuan, but Jun's travel agency sold them at 80 yuan. According to Shi, Jun had a good relationship with the World Phoenix Hotel, so that he got the ticket at a price as low as 45 yuan with decent seats. Before Hua left for the show, Jun's wife gave her 180 yuan to pay for four tickets, plus 10 yuan for her transportation. By coming to Jun's agency, the tourists could save (at most) 148 yuan per ticket, while Jun's agency could still earn 35 yuan per ticket while selling them at a discount.

In March 2013, in order to obtain the credentials for outbound tourism, Jun's travel agency became affiliated with the China International Travel Service Limited (CITS), as one of its service stations in Fenghuang. A few months later when I visited Jun, Jun handed me his new business card. His title was "Manager of the Local Hospitality Sector." I teased him, saying, "How come your business expands bigger, but your title gets smaller?"

2) Ma's Teahouse Gathering

Ma was the developer of Wucun, one of the Miao village tour sites. Ma split his time between Wucun and Tuo River Town, and the teahouse in the County Government Hotel lobby was one of his regular haunts in the *Gu Cheng*. One day in 2011, he invited me to his meeting at the teahouse with Jian, the developer of another Miao village tour site at Dongcun. The developer of the Heaven Dragon Valley later joined us. Among the three developers, only Jian was a non-local from Huaihua, a nearby city in Hunan.

According to Jian, the county Ethnic Affairs Bureau recently expressed an interest in forming a partnership with him to develop Dongcun village's tourism. Jian said to us, "I am not the least interested. If I could, I would not have bothered listening to them. I have always taken them out to dinners and given them gifts, and I thought that was enough."

Ma teased Jian, saying, "Sure, the government should have taken you out." Jian replied, "That is right. Shouldn't the government treat our outside investors well?!"

Ma responded, "You outsiders come to Fenghuang to do business. It is like picking vegetables in other people's garden. You left once done picking, and won't care about anything," Ma continued. "I am from here, so for me it is not just about making money. I care about my reputation as well. I told my fellow villagers in Wucun that if I make one yuan, I would like the village make 100 yuan. When I make one yuan, I still do much better than all the rest of the villagers who share 100 yuan."

Usually it was more challenging for the non-local developers to do business here. As a developer from outside of Fenghuang, Jian had a strained relationship with the villagers in Dongcun. Once one villager's water buffalo injured a tourist. Jian believed that the villager did it deliberately, as the tension between them had been escalating. It even mattered whether the developer was from the same village or not. Ma moved to Wucun through uxorilocal marriage.[4] He once went back to do tourism business in his natal village, and it did not last long. It was hard to deal with personal relationships where he had left for years. However, he had a much easier time doing business in Wucun. In Wucun, some villagers taught their children to follow tourists around, begging for money. The same happened in Longcun where the developer Lu was from outside Fenghuang. "I took take of it, and no more child beggars in Wucun," Ma said. "In Longcun, it's getting worse, and Lu could not get it under control."

Ma was concerned with the future of Wucun's tourism. Its best years were from 2006 to 2008, during which Ma made a profit of over two million yuan. He lost over one million in 2010, and expected to lose again in 2011. Ma attributed the decline of tourist visitations to the villagers' renovation of their old-style mud-brick compound houses into multi-story brick-and-tile residences. "Too many new houses now." Ma sighed, saying, "Every time I see loads of construction materials being transported into the village, my heart aches, and I say to myself, 'over, it's over!'" Ma recounted the story of one of the villagers who had been his good friend since they were young. The friend's house leaked badly when it rained, and he decided to renovate. Ma went to him, trying to talk him out of it. According to Ma, he said, "Look at my house. Is it livable? I actually want to ask you to loan me money to do the renovation." Ma said to me, "What he said was indeed the truth. How could I not let him?!"

Ma was thinking of changing his approach to tourism in Wucun. Rather than focusing on village tours, he was thinking of developing Peasant Family

Happiness (*Nong Jia Le*), a rural-themed resort with restaurant, guesthouse, and recreational agricultural activities such as harvesting fruit.[5] With an annual fee of 60,000 yuan, he could lease the mountainous land in Wucun for a term of 30 years, where he planned to build stilted wooden houses and raise free-range chickens and pigs. "No entrance ticket. I just want to attract tourists to come and spend money," Ma said. Ma asked the other two developers if they were interested in doing it together. Ma wondered if I could introduce to him any investors from Beijing. With a disability caused from falling off a cliff while cutting firewood many years earlier, the county government had honored him a "model" for disabled people due to his recent entrepreneurship. He planned to apply for a development grant from the China's Disabled Person's Federation, and he also wondered if I had any connections with the federation's headquarters in Beijing.

What Wucun faced was a dilemma commonly found for local Miao village tour sites. Teng, a Miao in her thirties, was one of the youngest officials in the county Ethnic Affairs Bureau. According to Teng, with the Bureau's sponsorship, Dongcun village used to receive an annual tourism development grant in the amount of 400,000 yuan from the provincial government. The provincial government ceased funding Dongcun in 2014, as the village did not pass its annual review due to "too many new houses." On one hand, Teng complained that the villagers were shortsighted and only cared about immediate benefits; on the other hand, she tried to put herself in their shoes, saying, "Making money is for a better life, and we can't blame them for wanting a toilet and hot shower, especially considering many of the old houses are dangerous to live in." Some villagers planned to renovate their old houses into two-story residences, and when the government allocated them money for cosmetic work on the new houses to conform with the traditional style for the sake of tourism, they instead used it to add an additional half story. "The village cadres in Dongcun lacked capability," Teng said. "Our bureau has switched to sponsor another Miao village now."

3) Pang Ge's Banquet

I met Pang Ge during the *Liu Ye Liu* in 2011 in Wucun. *Liu Ye Liu* was the sixth of June in the lunar calendar, an annual Miao festival day. I returned to the village from the nearby mountain slopes where the festival was held. It was past four o'clock, and I was looking for a late lunch. Next to the parking lot at the village entrance was a family restaurant. With the owner couple being away, their teenager son was left in charge. No other customers were there. The TV was on. I sat by a table eating instant noodles, the only option available at the restaurant. Pang Ge and several tour bus drivers were at another table playing poker, waiting for the tourists to be done with their tour. Pang Ge was a member of Ma's management team. Ma had placed Pang Ge in charge of the Wucun Miao Village Tour Reception Station in the *Gu Cheng*. Pang Ge offered me a ride in one of their tour buses back to Tuo River Town.

Pang Ge called me the next day, inviting me to join him and his staff for dinner in the *Gu Cheng* that night. It was Ma's treat, and Pang Ge reserved a private banquet room with a large table to seat twelve of us. There were ten tour guides and salesmen. Between late teens and early twenties, they were all local, except one from Guangxi and another from Sichuan. Like the majority of tour guides in Fenghuang, they were non-licensed. One boy had a nickname "*huoshao ji*" ("fire-roasted chicken"). Tour guide was his daytime job, and at night he performed fire stunts for the campfire-evening show at the World Phoenix Hotel. He was also a salesman, selling show tickets to tourists.

Over dinner, Pang Ge made a thankful toast to their hard work on the festival day in Wucun. Wucun received 570 tourists that day, and no tourist complaints were filed. One common complaint was about the price for its entrance ticket. For the local Miao village sites such as Wucun, between the undisclosed bottom line price (20 yuan or lower) and the official price printed on each ticket (around 100 yuan) there existed a wide range of actual prices that tourists ended up paying for. As *huoshao ji* put it, "It is like shopping at a clothing store. The customers with membership cards may pay a lower price than those who did not, as they have better *guanxi* [social network] with the store and received a greater discount. It is inevitable."

"To avoid complaints, the trick was to win sympathy," *huoshao ji* said. "We try to be apologetic. We tell the unhappy tourists their complaints will cost us our salary." He laughed, and continued, "Actually, Ma rarely deducted our salary, since it is already so low."

Pang Ge said to him, "You should rely on not Ma [for the salary], but the tourists [for the kickbacks], to make money. At least, you are better off than some of those *guo dao*."

The *guo dao* referred to the state-licensed tour guides. In some cases, the *guo dao* took tour groups from travel agencies or tourism companies, who charged them for each group at a certain rate per tourist.[6] They had to make sure that the tour groups they took spent enough on the tour. Otherwise, the kickbacks they made might end up being not even enough to cover what they paid to the travel agencies or tourism companies. After dinner, we went to a karaoke bar. A flyer was posted on the door that read "Janitor Wanted. Monthly Salary 800 Yuan." "That's even 200 yuan more than our salary!" one of the tour guides exclaimed.

The Entry Fee

Fenghuang made the headlines nationwide in 2002 for the county government's controversial deal with the YDCC. Eleven years later, Fenghuang found itself again at the center of national attention. On April 10, 2013, the county government implemented a new policy that required the payment of a

148 yuan fee to enter the *Gu Cheng* in Tuo River Town. The next day, the shop owners in the *Gu Cheng* went on strike to protest it. The fee allowed multiple entries per tourist to the *Gu Cheng* for two days, including the sites managed by the YDCC and the Nanhua Mountain leased to another developer.

To enforce the entry fee, numerous checkpoints were set up around the *Gu Cheng*. Entry to the *Gu Cheng* used to be free to tourists. Many *san ke* (independent tourists) chose not to pay to visit any sites leased to the developers. They instead wandered around in the *Gu Cheng*, enjoying its scenic views and riverside shops, restaurants, and bars. As tour group fees normally included the tickets to these sites, this new paid-entry requirement affected mostly the *san ke*, as well as the numerous traders and merchants in the *Gu Cheng* whose businesses were dependent on them. Since its implementation, the entry fee ignited fierce debate and protest. Many called for a boycott of Fenghuang and won support of the netizens nationwide (e.g., He and Wen 2013; Sun 2013).

Cai Long, the deputy head of the county government, claimed that their intention was to help protect the *Gu Cheng* through reducing the number of tourists. Many considered it "a lousy excuse," and accused the county government of "competing with the people for benefit." After the imposition of the entry fee, Jun's travel agency business dropped 40 percent. Even though Jun attributed it to the decreasing expenditure on tourist junkets resulting from President Xi Jinping's anti-corruption movement, he claimed that he didn't see the officially claimed benefits from the entry fee.

As pointed out by Hsing (2008: 60) in his study of the politics of China's urban land accumulation, parcels of land in China that are under the control of local authorities often serve as political resources that may be exchanged for favors or other resources such as investments in industry or urban infrastructure. This situation exemplifies Harvey's (1989) theorization that urban governance in late capitalism is marked by a shift from managerialism to entrepreneurialism. Fenghuang's case supports these observations. The county government used its land as 49 percent equity share in partnering with developers, to impose access restrictions through the new entry fee requirement in the *Gu Cheng*, and took a 60 yuan share from each ticket sale. As such, the county government assumed a dual role as both market regulator and market participant. Its entrepreneurial interests were mixed with its public interest as noted elsewhere by Sargeson (2004: 644), which challenged its "moral capacity" (Hsing 2008: 58) as the legitimate regulator and protector of public interest.

1) Yao's Opinion

Amid the media's massive coverage of Fenghuang's new entry fee, a Beijing based Agence France-Presse (AFP) reporter contacted me for my opinion in late April 2013. I was then finishing up teaching for that semester in Michigan. A few weeks later, I flew to Beijing. I called San Ge. His rental van business was not as good since the implementation of the entry fee. He complained that he made no more than when he was a migrant factory worker in Hangzhou. Jing's restaurant had fewer customers as well. Jing did not need San Ge to help as often, let alone hire additional hands.

Shortly after, I arrived at Fenghuang. After a six-hour ride from Changsha, I got off the long-distance shuttle, and grabbed a taxi heading toward the *Gu Cheng*. There were fewer *san ke* in the streets. The taxi driver said, "Even though it is the peak season right now, the tourists are only as many as on weekends in non-peak season." I noticed several of the big silver jewelry stores were closed. "More shops will be closed by the end of year, once their rental leases expire," he asserted.

I visited Yao in his office the next day. Yao, the vice director of the county Statistics Bureau, was among the few local officials who I interviewed in 2005 and we had become friends since then. When I arrived, he was reviewing a research report by a student group from the Shanghai University of Finance and Economics. During their research trip in Fenghuang, they interviewed Yao, and they had later sent their draft report to Yao for comments. Since its tourism boom, Fenghuang had received numerous scholars and college student research groups from all over China.

"Come and take a look at this. What do you think?" Yao said as he gave his seat to me. The report investigated the impact of the entry fee policy, with Yao's comments on the margins. I asked Yao about the seemingly contradictory replies from the county government regarding the entry fee's impact on the tourist numbers. On one hand, the county government claimed that this policy was to decrease the number of tourists to better protect the *Gu Cheng*; on the other hand, while responding to the traders' concerns about their declining business, it instead claimed that the tourist numbers actually increased during implementation.

Yao paused for a few seconds, and said, "The government has to say whatever is suitable to the different situations." From his pocket, he took out a meeting notebook. He opened one page, and pointed to the meeting notes he took the day before. It read: "1–13 August, average 11,500 entry tickets sold daily; 1–8 July, average 18,000–23,000 tourists per day who checked-in (based on the monitoring points set up at 75 lodging places)." He explained, "These are close to the target numbers the government desires. There are about 40,000 beds in Tuo River Town right now, and the government wants to eliminate over 10,000 beds from the market through implementing the entry fee."

According to Yao, a journalist wrote an article regarding the impacts of the entry fee during the first three months, which focused on how much the fee had damaged small businesses. Concerned with more protests, the head

of the county government had mentioned the article at this meeting to the attending officials including Yao, to urge them to be alert to the public opinion
influenced by mass media. Yao was optimistic. He said to me, "The entry fee
severely impacted the local traders who had relied on the *san ke* as their main
customers, however it is much better now than it was back in April. Life is
always hard. Ordinary people have little power to change the reality, but they
are resourceful in adjusting to it."

However, as the system allowed little else in the way of opposition, ordinary people's pragmatic adaptation to reality was simply tactical wisdom, and
by no means implied normative consent to those realities. During my trip in
Fenghuang for the remaining summer of 2013, I visited with my long-term
informants and friends and made new ones, many of whom were doing small
tourist businesses in the *Gu Cheng*, and witnessed their struggles under the
new entry fee policy.

2) The Traders

A few days after my visit with Yao, I took Liang and her husband Hong
out for lunch. While Liang was a camerawoman, Hong worked as a self-
employed construction contractor in town. Hong talked about the strike
back in April. The county government did not use force against the crowd
of protesters during the demonstration, but afterward the police came to
the homes of individual participants at night. Some were detained for ques-
tioning, including Hong's friend Yong. Hong recalled that Yong was not
concerned before the police came to him. He joked with Hong, saying, "I
just moved into a new house—they can't find me." After the police took
Yong away, his wife turned to an acquaintance for help, an individual who
was a colleague of the police officer who detained Yong. The acquaintance
assured her that the police took Yong only for a talk, and they wouldn't beat
him. "They won't beat people. They don't want to escalate conflicts," Hong
said. "They are good at strategically cracking down by covertly targeting
the protestors one by one."

Our lunch was at the "Big Wok and Big Stove" (*Da Guo Da Zhao*),
a restaurant recommended by Liang where her friend Zhao worked. It
specialized in firewood-cooked hot-pot fish, a popular cuisine from the
restaurant owner's hometown in Yiyang (another city in Hunan). The
restaurant was a stone's throw away from the Rainbow Bridge, a good
location for business. Zhao, her husband, and their five-year-old son had
moved from Yiyang just before the implementation of the entry fee to
work at the restaurant, which was owned by the brother-in-law of Zhao's
biao ge (maternal cousin).

With the implementation of the entry fee policy, the restaurant business quickly declined. While Zhao's husband looked for other temporary jobs, Zhao purchased a set of Miao costumes with elaborate embroidery and head-dresses. In the morning, she dressed up with the costume and makeup, and placed herself in tourists' favorite photograph spots. For those interested in taking a picture with her, she asked for 5 yuan. Right before noon, she hurried back to the restaurant, where she took orders, served dishes, cleaned tables, and swept the floor. In the evening, if the restaurant was not busy, she put the costume back on, and with a quick makeup touch-up, she was ready to return to the streets for her picture business. There she made friends with Liang, who frequented the same spots as a camerawoman.

Xia and her husband Shun ran a riverside shop selling CDs and "hand-woven" scarfs. Shun told me that because of the entry fee, their shop revenue was only one-third of the year before. Shun questioned the county government's motivation. "They claimed that it was to foster a high-end tourist market. How can it be?!" Shun continued, "Most of the tourists here are tour groups. They might pay 80 yuan for a tour package if they are from Zhangjiajie, or 200 yuan if from Changsha, which then had to cover the costs for their transportation, lodging, and entry fee. If they were rich, why would they bother joining tour groups? If they join tour groups, they are unlikely to have tens of thousands to spend here."

Xia's father, Hai, recently moved with his wife to the *Gu Cheng* from Wucun. There, they rented a riverside property to run a guesthouse. Hai told me that a guesthouse owner on another street in the *Gu Cheng* had taken pesticide to commit suicide, because he started his business with a bank loan right before the sudden implementation of the entry fee and subsequently lost his investment. However, Hai said that his business, along with the other guesthouses on the same street, was not affected too badly. No checkpoint was set up on this street lined with guesthouses. On the other side of this street was a temporary long-distance bus station, where the arriving tourists could walk along this street to the Rainbow Bridge without paying the entry fee. "The government turned a blind eye to it. Otherwise, all the guesthouses in this street would have a hard time to survive," Hai said.

The county government made compromises with other small businesses in the *Gu Cheng*. To appease the local peasants who ran boat sightseeing businesses, the county government agreed to compensate each of them with 50,000 yuan annually. These local boatmen were not allowed to enter the scenic part of the river monopolized by the YDCC. They docked their boats by the lower stream, and walked to the scenic spots to seek tourist costumers. "We liked the *san ke* a lot more [than tour groups]. They are not bound to take the YDCC's boats, and ours are cheaper," one of them said. "We rely on the *san ke* for business."

Yuan was the manager of a silver jewelry shop in the *Gu Cheng*. It was a local family business that for four generations practiced making traditional handcrafted Miao-style silver jewelry. Yuan's shop did not develop partnerships with tour guides, who normally demanded a kickback in the amount of 50 percent of the relevant sale. Yuan's wife, working as one of the two saleswomen in the shop, said to me, "The kickbacks are solid, which would have forced us to reduce our cost on the stuff sold to the customers. We are not willing to reduce the quality of our products." The shop's customers used to be local residents, and the shop earned its good reputation by word of mouth. According to Yuan's wife, most of the shops in the *Gu Cheng* were run by itinerant merchants, and they switched owners frequently, with an average turnaround time of three years. She pointed at a store across the street, saying, "It was a silver jewelry shop last year, but an embroidery shop now."

As those who shopped here were largely the *san ke*, the sales of Yuan's shop dropped since the entry fee started. Yuan's mother, the registered owner of the shop, was the head of the county Chamber of Commerce. Yuan said that she complained to the county government, and the county government offered rent reduction, as the shop front was a government property. "Many small businesses were closed," Yuan said. "On the surface, it results from market competition; but at the core, it was intentional through policy." Yuan told me what he believed was the county government's motivation: the imposed entry fee led to fewer tourists, fewer tourists meant less business, and less business made the real estate value drop, so that the government could seize more land at a cheaper price for development projects.

Land as "Super Wealth"

What Yuan believed was not groundless. Land values had increased with the tourism boom, and the county government had keen interests in it. Political authorities in China often employ the critique that collective ownership promotes egalitarianism and leads to unscientific and undercapitalized farming, and they therefore advocate liberal reforms to rural land rights (Sargeson 2004: 642). In Fenghuang, aside from the direct linkage of joint ventures and subcontracting, the local government restructured to facilitate the influx and expansion of outside capital, which involved the requisitioning of farmland for commercial use with low compensations for those with use-rights.

As Hsing (2008: 63) has pointed out, the local governments across China adopt the doctrines of rationality and efficiency embodied in capitalist land use planning that "urban planning is fundamentally about realizing the exchange value of land on the market and about allocating land in market-efficient ways." Changwen Yan, the County Communist Party Secretary of

Fenghuang, spoke at a County Biannual Meeting of Economic Work (Yan 2013): "If the land is in the hands of peasants, its value lies just in planting crops; if the land is in the hands of the government, through planning, investment, and construction, its value increases dramatically. It becomes a finance asset, becomes super wealth, becomes the platform for development, and becomes the carrier for urbanization."

1) Land Requisition

In contemporary China, the peasant households have contractual use rights to farmland, but ownership rights lodge with the village collective unless the government requisitions the land (Sargeson 2004: 637–638). There are two types of requisition by the government, one for non-profit purpose, and the other for commercial use. National laws stipulate that villagers' collectively owned land could only be requisitioned officially for the public interest (Sargeson 2004: 644). For example, it is legitimate when the construction of highway, hospital, or school necessitates it. Nevertheless, requisitioning farmland for commercial use was far more common in Fenghuang.

As elsewhere in China, the local peasants have transferable use rights while the village collectively owns the land. While leaving to pursue non-agricultural income opportunities, villagers were often willing to transfer their use rights to private developers through long-term rental contracts. In a village in the outskirts of Tuo River Town, some villagers had a vacation in Beijing. The trip was organized by the village committee, and paid with the money made from the village's farmland transactions. Land transaction in this village required written proof from the village committee, who charged a fee of 3 percent of the sale price. Due to potential benefits from farmland transactions, there was unusually fierce competition for the position of village head in those villages close to tourism attractions. In one case, one candidate, who made money as a construction contractor, made a promise to pay for all the villagers' social security for two years. He won.

The county government often played the role of middleman for developers in land transactions for commercial use. Yao's relative had to move the ancestral grave, as the land was being expropriated for real estate construction. He went to the developer's office to sign the agreement, and noticed that on the agreement the other party was not the developer, but the county government. His compensation was in the amount of 18,000 yuan. He counteroffered for 6,000 yuan more, which was accepted by the developer without much difficulty. In retrospect, he told Yao that he should have asked for more. Yao complained that such a procedure was not transparent, and they wondered if the county government or the developer benefited from the underpaid compensation.

2) Government Relocation

Much of the land in the *Gu Cheng* was leased to private developers, including even the land on which the county government office building was located. This five-story building was the home for most of the county bureaus, and with the land leased to private developers, the county government planned to build a new office building in the Red Flag New District outside the *Gu Cheng*. However, as the deadline to vacate their old building was approaching, the new office building had not even received the construction approval. Since President Xi Jinping took office, such construction projects had been greatly discouraged. The county government then had to look for temporary office space, and various bureaus ended up scattering all over town wherever suitable rental space was available.

The move was completed within a month by the end of March in 2014. The county Ethnic Affairs Bureau moved into a building in the heart of a commercial district, surrounded by retail stores with loud speakers announcing promotional sales. After our lunch at a nearby restaurant, Teng showed me to her new office. Passing through the dry cleaner on the ground floor and climbing up the stairs, we came to the bureau's offices on the second floor. Teng shared her office with a colleague. Music came from below, mixed with the sounds of kids playing. "Is it a kindergarten?" I asked. "No. Those were the kiddie rides by the store entrances," Teng said. The room was dark, despite the bright sunshine outside. "Two of my plants died," said Teng as she pointed to the row of potted plants by the window.

The layout of her office was similar to a hotel room. According to Teng, this building used to belong the county Agricultural Machinery Administration Bureau, who sold it to a private developer to run a hotel. The hotel was closed down soon due to lack of business. Its rooms had been left unused, until the county government rented it from the developer, and used it as temporary office space for its seven bureaus, with a dining facility set up for them. Teng envied the county Statistics Bureau that moved into the office building of the county Finance Bureau in another location, which had the best canteen. "You should visit Yao and try their canteen," she said, " it is as good as the restaurant we went for lunch today."

The next day, I visited Yao. Yao's and Teng's offices used to be next door to each other, as the two bureaus shared the same floor in the old government building. After the move, the county Statistics Bureau took the seventh floor of the county Finance Bureau's office building at the other side of town. With a tin roof, this entire top floor was an add-on. The building sat in a courtyard, secluded from the streets. It had an elevator, and a gym equipped with a treadmill, pool table, and Ping-Pong table. Yao's office was at the end of a hallway that led to a spacious open rooftop. On the rooftop, Yao showed me the peppers, grapes, and other vegetables and fruits he planted in big pots. He enjoyed his new office, saying, "We were lucky to end up moving here, thanks to our bureau head who had a good relationship with the county head." Later that summer, I was caught in heavy rain one day and sought shelter under the old government building's eaves. A middle-aged man was there before me. We chatted while waiting for the rain to stop. He was a staff from the county Animal Husbandry Bureau. "It was a lot more convenient when most of the bureaus were here in the same building," he said to me. "Since the move, not only can't the ordinary people find the bureaus to which they need pay visits, but also the government employees like me don't have much of a clue where many of the other bureaus are." Behind us was the empty old government building. It had been locked down, except being briefly used as a flood shelter in May.

NOTES

1. Rack's research (2005) focuses on the Miao in West Hunan, but it is worth noting that the ethnography was conducted in the city of Jishou and the surrounding lowland areas, due to the conditions of her research permission that prevented her from working in the highland rural regions where the Miao are concentrated.

2. Through two millennia of empire-building, the Han Chinese state expanded from a relatively small core area in present-day northwest China to encompass territories even greater than the current borders of the People's Republic of China (PRC). In Swain's discussion of local identities and transnational linkages in Yunan, she included a review on China's frontier studies (2002:182). The idea of a frontier zone was proposed by Aris (1992:13) in his writing about Tibetan borderlands, as well as by Gaubatz (1996:13) in his study of frontier cities and urban transformations in China. Major anthropological studies on ethnicity in China's southwest frontier zones have heavily drawn ethnographic data from Guizhou (e.g., Oakes 1998; Schein 2000), Sichuan (e.g., Harrell 2001), Yunan (e.g., Swain 2002), and Guangxi (e.g., Litzinger 2000).

3. The exchange rate of US dollars to Chinese yuan was about 1:8 in the early 2000s.

4. While patrilocal marriage predominates in rural China, uxorilocal marriage, in which the husband moves to live with his wife's family, has a long history.

5. For a detailed discussion on Peasant Family Happiness (*Nong Jia Le*), see Chio (2014: 95–97).

6. The rate was based on the estimation of the expenditure of a particular tour group. For example, a higher rate for tour groups from Guangdong than those from Guizhou, as the former tended to be more generous than the latter.

Chapter Two

"Scaling-Up"

Mcdonaldizing Village Tours

After an overnight train and six-hour bus ride from Beijing, I found myself at the long-distance station in the heart of the new district of Tuo River Town.[1] By the road with my backpack and suitcase, I was trying to spot a taxi. A woman at a roadside restaurant shouted at me, "No taxi today." I was about to give up when a taxi pulled up. The taxi driver told me that most of them were on strike that day, to protest the county government taking back their ownership rights to the vehicles. About a dozen (out of a total of over a hundred) taxies were doing business that day. On our way to the *Gu Cheng*, a government staff stopped us and took the name of the taxi driver. She told him that the government wanted to identify those who did not participate in the strike, and might reward them with priority when renewing their operation licenses.

The taxi dropped me at the County Government Hotel. It was in the same compound with the county government office building. I would be staying there for a few nights before leaving for the countryside. The courtyard of this compound was used as a parking lot for tourists. Among the three authorized commercial parking lots, this was the closest to the *Gu Cheng* downtown. It was packed with tour buses and sedan cars with non-local license plates. It was mid-June in 2011, still a month away from the summer peak season. Right below my window on the third floor parked the tour buses. Loud noises woke me up early in the morning. It came from the running engines and air conditioners, mixed with tour guides herding their tour groups and loading them onto the buses. Many of them were heading to Fenghuang's countryside for a one-day Miao village tour.

"MCDONALDIZATION" AND THE POWER OF SCALE

In this chapter, I explore the role of tourism mediators, particularly tour operators, as the primary agents of change in the growth of packaged mass tours in Fenghuang, paying attention to "the power of scale" (Bodley 2003). Scale refers to the absolute size of population, economic enterprises, markets, armies, cities, or anything that affects the well-being of people. The power of scale refers to the reality that scale increases can be expected to mathematically produce disproportionate concentrations of power for those at the very top of any hierarchy in any power domain, while the costs of growth are likely to be socialized or borne by society at large. It suggests that scale gives elites a powerful incentive for growth, and offers insights to the "McDonaldization" process (Ritzer 1993) in tourism industry.

McDonaldization has been proposed as "the process by which the principles of the fast-food restaurant (efficiency, calculability, predictability, and control) are coming to dominate more and more sectors of American society as well as the rest of the world" (Ritzer 2010a: 4). Exemplifying the contemporary rationalization process, the McDonaldization thesis has been used to interpret broad trends and tendencies across the tourism industry (Bryman 1998, 1999; Ritzer 1998; Ritzer and Liska 1997; Weaver 2005).

Despite the benefits (e.g., efficiency, predictability, standardization, and affordability) for corporations and consumers, the process of McDonaldization often brings with it problematic consequences (e.g., dehumanization, alienation, and poor-quality work), hence the "irrationality of the rationality" (Ritzer 2010b). The existing limited discussion on the negative impacts of the McDonaldization process in the tourism industry have largely centered on the production dimension from the perspective of the tourism corporations (e.g, Weaver 2005), rather than on the consumption dimension from the experience of the toured and tourists.

Furthermore, insensitive to variations across different countries and cultures, Ritzer's McDonaldization thesis also somewhat neglects the cultural side of the McDonald's phenomenon (Kellner 1999; Watson 1997). While Kirshenblatt-Gimblett states that standardization is "part and parcel of the economies of scale that high-volume tourism requires" (1998:152), there exists "standardized diversity" as Di Giovine (2009:166) calls it, to emphasize the balance between standardization and differentiation that the travel industry practices. As Salazar points out, much scholarly endeavor has been devoted to demonstrate that "the local is not and never was the passive, bounded, and homogeneous entity it is frequently assumed to be" (2005: 629). Through looking at the key roles that local tour guides play in mediating the tension between ongoing processes of global standardization and local differentiation, Salazar (2005, 2010) argues that globalization and localization are ultimately intertwined through the process of "glocalization" in the context of tourism. As a relatively recently conceived interpretation of contemporary society, innovative but ultimately incomplete (Smart 1999; Weaver 2005), the McDonaldization thesis deserves more consideration in anthropological tourism studies to examine the role of tour operators in shaping the experiences of both the toured and tourists in tourism destinations. Focusing on the scale issue in the process of McDonaldization, this chapter draws on the power-elite hypothesis (Bodley 1999, 2001) that when growth in the scale of businesses and regional economies is an elite-directed process, it may concentrate social power and diffuse costs, to illustrate the "irrationality" (negative impacts) of the "rationalized" McDonaldization process in the tourism industry in Fenghuang.

PROMOTING COUNTRYSIDE TOURS

In China, as an important strategy for poverty alleviation, one significant trend in tourism is the *"xiangcun you"* ("countryside tours"). In 2006, China's National Administration of Tourism announced the tourism theme of the year as *"xiangcun you,"* with the promotion slogan "new countryside, new tourism, new experience, and new trend." In 2010, China's National Tourism Administration and the Ministry of Agriculture signed an agreement to jointly propel the development of *"xiangcun you,"* to establish a list of model counties/towns/villages through three to five years of effort. *"Xiangcun you"* catered to urban people's desire of escaping from the city anxiety and pollution to more pleasant country living for recreation on a budget.

In the context of nationwide promotion of *"xiangcun you,"* Fenghuang enthusiastically embraced this strategy. For the county government, *"xiangcun you"* seemed to be a perfect solution to two major problems. Ideally, at first, it would greatly relieve the pressure concentrated on its urban center Tuo River Town; second, it would contribute to correcting the uneven distribution of economic benefits between its urban and rural communities. *"Xiangcun you"* was favorably perceived not only by the county government, but also by the local residents who were eager to have a fair share of the tour market through developing village tours.

"In Fenghuang, *'xiangcun you'* is village tours, especially Miao village tours," the deputy director of the county Tourism Bureau explained. According to her, six village-tour sites were officially opened by the summer of 2011, with another dozen running for trial. Fenghuang's tour market lacked regulations, and she thought that the high kickback was the root of problems.

She told me that the bureau's response to the problems was to "let it be": since it was difficult for the government to forcefully close the illegal village tour sites, the best solution was to leave them to the market, which would eliminate them through competition. "Doing less is actually solving more," she said. "It is troublesome if conflicts were intensified between the government and local villagers."

"Survival of the fittest" sounded like a rational choice to eliminate the illegal tour sites without causing the government much trouble. But whose trouble might it be then? Many of these tour sites paid no taxes and bought no insurance. With the high rate of tax evasion, the county head once said, "Fenghuang tourism gets its people rich, but not the county." Without official inspection and approval, those self-made crude tourist facilities raised serious safety concerns. Without issuance, it was up to the tour developers to determine the amount of compensation for any accidents. In July 2010, a tourist drowned while swimming in the "Ancient Monster Pond" in Wucun village. He was from Shandong, and was then a college student in a university in Hunan. Ma told me that he was fined 100,000 yuan by the county government. In addition, he paid 200,000 yuan to the student's family as compensation. Ma complained that he had to pay it out of his own pocket. He tried to settle it with that family for as less money as possible. He said to them, "I have a good lawyer. You either accept my offer, or we go to court and you might not even get as much."

Lu, the developer of Longcun village, once commented on these illegal tour sites, saying, "Any Miao village could just invest 20,000 yuan, nail a few boards to build a gate, and start selling entrance tickets. The next day, its villagers go to beat drums in the streets to promote it. No business license! Nothing!" I asked Lu, "How about the county Tourism Bureau? Do they do much at all?" Lu replied, "Once in a while they tell us 'Big wind is on its way. Close your site,' 'It's about to rain. Close your site,' or 'It's market day in Shanjiang. Close your site. Otherwise, there will be traffic jam.' And I would say traffic jam has nothing to do with me. It is your government's business. Even if it lasts till tomorrow morning, it has nothing to do with me!" He complained that the local government including the county Tourism Bureau did nothing to regulate the tour market, and it was difficult for the developers like him to run their businesses. He said, "They don't realize that the ultimate victims are the Fenghuang people themselves. The victims are not we the developers as we can pull out our investments and leave, nor the tourists as they could afford to waste 100 yuan on an entrance ticket."

Ru's husband, an official from the county Water Conservancy Bureau, offered further insight into why the local government did not try hard enough to regulate the tour market. He said, "Many villages that established tours are not licensed. But no one dare shut it down. If the law enforcement people go to force the closure, the villagers would say 'we are doing it in our own village. We ordinary people just try to increase income. We are not stealing. How wrong could it be? Why do we need to apply for any license?' The government head of course do not want to press too much from above to result in intensified conflict, which will affect their performance evaluation during their term." Some had another explanation: the problematic village tour businesses had strong *guanxi* with the local political officials. *Guanxi*, as a form of social capital that involves reciprocal responsibilities based on familial and clan relationships (Lew and Wong 2004), played an important role in the local tourism market. In Fenghuang, as elsewhere in China (Lew 2007), the local officials' decisions were based more on *guanxi* business relationships than formal business plans. Some officials even had a share in those village tour businesses.

A manager from the Phoenix Ancient Town Tourism Co. Ltd (PATT) pointed out another problem with Fenghuang's *"xiangcun you."* According to him, the PATT was the *datou* (literally "big head," the single dominant power) in Fenghuang between 2002 to 2005, but now the market share of *"xiangcun you"* has increased a lot, emerging as the PATT's major competitor. "It is an unhealthy competition," he asserted. "They sell the entrance ticket at 148 yuan, the majority of which flows back to the tour guides and agencies, therefore the actual amount remaining in the village is only 5–10 yuan per ticket." He further commented that the price for a tour package was low, attracting many middle-low income tourists from around west Hunan and east Guizhou. With consumption potential per tour group decreasing, tour packages then tried to squeeze more activities and/or sites into a tight schedule. He believed it was not beneficial to local communities and Fenghuang's tourism overall. He used data to support his point: in 2009, village tours received 941,000 tourists, and made ticket sales of 49.43 million yuan. Thus, the number of tourists increased 54 percent; however, tourism income only increased 11 percent in 2009. Nevertheless, the PATT added one *"xiangcun you"* site (Longcun village) in 2008, as a response to the new trend in Fenghuang.

VILLAGE TOURS

Since the promotion of countryside tours, the main streetscape in the *Gu Cheng* had become dotted with private tour companies that offered various tour packages. The tourists who desired village tours normally purchased tour packages from tour operators who provided transportation. Otherwise, it was impractical (if not impossible) for those who were not backpackers to do a one-day tour in a remote village on their own. With the popularity of *"xiang-cun you"* penetrating the countryside previously not influenced by tourism, village tours were having profound impacts on Fenghuang's tourism market and its rural communities. In the following, I discuss three village tour sites including Wucun, Huangcun, and Longcun, which were developed by either local elites or outside investors. The discussion demonstrates the experiences of both the tourists and the local villagers, reveals the role played by tour developers and operators, and explores the dynamics of the global process of "McDonaldization" in a local situation of mass tourism.

The "Over-Packaged"

I met a middle-aged tourist with her daughter at a restaurant in the *Gu Cheng.* We had to share one table for lunch, as it was over-crowded. They were from Zhuzhou, a city in Hunan. The trip was a gift for her daughter's middle school graduation. Before their trip, they went online and purchased a tour package. They felt so unwelcomed at a guesthouse that they had to move out after a one-night stay. Many guesthouse owners developed business "webs" with the private tour developers who operated individual tour sites. The guesthouse owners earned kickbacks from these developers for entrance tickets or tour packages that they sold to the tourists who stayed at their guesthouses. Since the mother and daughter tourists purchased the tour package elsewhere, the guesthouse owner found them less profitable than other tourists who were looking for lodging and a tour package at the same time. They just returned from their village tour. They joined a tour group to Wucun that day, hoping for an authentic experience of Miao culture, but were disappointed, as it was not at all what they had expected.

A few days later, I visited Wucun. Located in Luocaojing Township in the southwest of Fenghuang, Wucun had a total population of around 2000, the majority of which were the Miao. It had been a "model" Miao village tour site promoted by the local government. In my previous work, I discussed in detail the village and its tourism development directed by the village elites (see Feng 2007a for details). It had been four years since my last visit in 2006. During a one-hour trip, a minibus took me from Tuo River Town to Ala Town, and then a three-wheeled motorcycle "taxi" got me directly to my destination. Riding in the old dusty minibus with local people and sometimes their chickens was quite more enjoyable than the air-conditioned tour shuttle. There was no direct public transportation from Tuo River Town to Wucun. Not many tourists were willing to take the trouble or simply didn't know how to travel to the countryside on their own.

The village tour started with the "three roadblocks" welcoming ceremony[2] at the village entrance where rice liquor was proffered, songs sung, and drums played by the tour operator's employees. Then the tourists entered into a big cement-paved parking lot, with a two-story wooden building encircling it. It hosted the village conference room, the office of Ma, and a two-story display space for tourists. The display house included a large "divine" room on the ground floor devoted to the so-labeled "Miao deities," and the upstairs was for the exhibition of the Miao's traditional farming tools and household utensils. The tour guide led the tourists into the "divine" room, where poles painted with colorful images featuring unidentifiable totemic beings, and twelve paintings featuring the deities hung side by side on the wall. There was a long table under the paintings, on top of which were placed twelve plastic baskets corresponding to each deity. Individual tourists were instructed to find his/her matching protector based on their birth month. The guides' dispassionate voice magnified through the speaker introduced the deities and their power to bring each tourist good fortune if "respect" were shown by putting money in the basket. Enough time was ensured by the guides for the tourists' stay, and then a quick walk-through of the upstairs exhibition ended the first part of the tour.

A narrow passageway connected the parking lot with the circular performance ground. Led by their tour guides, groups of tourists went through the vendors lined up on both sides of the passageway. They came to watch a show of Miao dance, music, and games, constructed as the central tour experience. The performance, choreographed by outside professionals commissioned by Ma, had not changed much for the past four years, including a water buffalo dance showing the daily life in the field, the market dance revealing courting among Miao young people, the stunts of climbing the knife ladder and walking through fire, and the interactive activities of inviting tourists to join the performers for the bamboo-stick dance and a singing contest. But at the very beginning of the show, an auction section was added. The show host displayed several Chinese watercolor paintings by artists from around China for auction.

The show lasted for about an hour. Almost immediately after the show, the tourists were herded through another passageway of the performance ground to the dining area. Long rectangular tables and benches were placed together to make a giant dining table, where rice and dishes were already served. With the land leased from his fellow villagers, Ma recently built this dining facility, and had his wife worked there. The cost of the meal was included in the package fee the tourists paid to the tour agencies, from which Ma got 10 yuan per tourist. Thus, Ma not only managed exclusively the tourists' "consumption *of* the journey" (the "material" of attractions at destinations), but also started controlling the tourists' "consumption *on* the journey" (the "material" serving of such daily consumption needs as eating) (Wang 2006: 71–72). The profit from the tourist dining in the past year was 300,000 yuan.

Travel agencies and tourism developers' control over tourists and restriction of access are common in China's villages and small towns open for tourism (Svensson 2010). The tourists in Wucun were like the commodities in
an assembly line: they were being furnished with a standardized mass manufactured tour experience along the rigidly constructed tour route. They had
little time to visit the nearby vendors and stalls, or to dine at their own choice
of the few family restaurants and food stands, or to reach beyond the staged
front of the village centered on the performance ground to interact with other
villagers. The family restaurant at which I once dined was closed. Another
stayed open, with little business. It was where I met Pang Ge later. Most of
the time, it became a gathering place for tour bus drivers to chat, watch TV,
and play poker while their tour groups were touring the village.

Hai, the director of the village performance troupe, was one of my key informants in Wucun since 2005. He managed a souvenir stand with his wife at
that time before their moving to Tuo River Town. He told me that their small
business was getting harder to do, with the tourists always coming and going
in a hurry. I visited a Miao woman in her old wooden stilted house. She was
in her seventies, and her house was away from the standard tour route. Occasionally, when a few *san ke* found their way there, she demonstrated how to
make cotton strings and weave on her wooden weaving machine, and showed
them the cloth and clothes she had made that way. This was in contrast with
the display house in the village parking lot complex: one was stiff, distant,
and staged; the other was alive, interactive, and of everyday life. She did not
demand, but would accept, any money from the tourists who visited her. I
asked her why she didn't join Ma's village tour business for a much larger
and steady number of tourists. She replied that Ma was not interested.

Close ties between the tour agencies and village tour developers excluded
the majority of the villagers from the tourism space and, therefore, enabled
their control of tourists through producing and promoting "McDonaldized"
tour experiences. While providing poor-quality tour experiences to tourists,
the "over-packaged" village tour enhanced the tourism operators' profits,
and left limited benefits to fewer villagers. This contrast was demonstrated in
the image of Ma's brand-new Volkswagen Passat sedan in the village, seemingly out of place. But these days, Ma had his new worries as he constantly
expressed during our several conversations. He had to think hard about the
new ways to make the village tour "sustainable." He complained that one of
his big "headaches" was the new brick and cement houses replacing the old
mud-block traditional Miao-style houses,[3] as mentioned in the last chapter.

Ping was the village Communist Party Secretary. Ma had him manage
the daily tour operation within the village, and paid him a monthly salary.
I asked Ping whether the tourists liked their tour experience in Wucun. He

said that their most common complaint was the village being too "modern." Hai once heard a tourist saying, "Are you sure this is a Miao village? Your Miao village doesn't look much different from where I am from. What's the point for me to tour here?!" Ping told me that Ma had sent his teenaged son to learn the Miao stunts to perform in the show. As all the shows at the local Miao village tour sites were nearly identical, Ma wanted to add special performances. "Ma thought that since our houses are no longer good, we need good performances," Ping said. "But if Ma sent someone else to learn, he may no longer be interested in staying in Wucun to perform; or, he can demand a high salary."

The "Abandoned"

Located in Ala Town, Huangcun is close to Wucun. Much smaller than Wucuan, it originally had 97 households with a population of 475. Different from other village tours in Fenghuang, Huangcun's tourism had little to do with ethnic minority culture, since the villagers were predominantly Han. However, the village's historic rampart[4] was one of the eight attractions that the county government leased to the YDCC in 2001, and it had been one of the well-known sites managed by the PATT. As the PATT monopolized the profit, the villagers fought with the PATT over the village's tourism resource. The conflict was much intensified when the county government ordered the villagers to move out at the request of the PATT to restore the entire village. In 2005, one-third of the households refused to move out at the given rate of compensation. The villagers continued to sell their own tickets and showed tourists around the village and sometimes sneaked up onto the rampart. During my stay in 2006, the villagers were somewhat successful in resisting the pressure from both the PATT and the county government, and collectively developed the village tour, apart from the PATT's rampart tour (see Feng 2007a for details).

Before long, Huangcun became an "abandoned" tour site. On January 8, 2007, the PATT closed its tour at the village. At the PATT's headquarters in Tuo River Town, I interviewed Manager Zhang, who had worked at Huangcun in 2004, and was the associate director of the site back then. He explained the closure was due to the resettlement issue. "The social environment was messy there," Zhang blamed the county government for causing the situation. "The government did a poor job in moving and resettling the villagers." According to Zhang, the government did not plan it carefully, and ran out of compensation money quickly. Zhang believed that the villagers were willing to move out, and the problem was due to the government's failure in distributing the compensation well. He said, "The villagers want us to continue business there. Since there are no tourists now, their income is affected negatively very much. They used to make decent cash to help household income by solely relying on selling deep-fried river crab and boiled corn on the cob." According to Zhang, the PATT did not have any plan right then, not until the resettlement issue was resolved.

"We don't have any problem with the PATT. It is mainly the government," one villager told me. It in a way confirmed what Manager Zhang said about the villagers wanting the PATT to continue business there, but did not necessarily mean that there was no conflict between the villagers and the PATT. There had always been quarrels between the PATT employees who worked at the rampart and the villagers. From the PATT's perspective, the villagers did make it very difficult to maintain its regular business. Back when the PATT managed the rampart, some villagers added additional floors on top of their flat house to reach the top of the rampart. They then made the top floor open toward the rampart trail to sell small goods to the tourists. Responding to the PATT's negative reactions, the villagers argued: "We are not taking your rampart space, but only doing business in our own house."

One incident might further elaborate what Manager Zhang called the "messy social environment." A villager who lived near the village entrance gate took some coal cinders to dump them, and dropped some on the ground by the gate on his way out. A PATT's tour guide was upset at him for making the site dirty, and they started a quarrel. Other villagers joined and stood by their fellow villager. The guide then gathered two cars of people from Tuo River Town to the village. According to the villagers, she was from Tuo River Town, and had a relative working at the county Police Bureau. They said that it was common for the local officials to arrange for their relatives to work for the PATT and other tour operators. The conflict soon escalated. The villagers cut the cable to disconnect the PATT's power supply line that ran through the village. The next day, when a tour group from the provincial education system arrived to visit the rampart, the villagers blocked the entrance. The county government had to send a mediation team to settle the intensified conflict.

Tao was among those two-thirds who accepted the compensation and moved out. She received a three-floor unit by the main road at the center of Ala Town. However, she did not have the title of the property. "The property was worth 180,000 yuan six years ago, which is only enough for one of its rooms now," Tao said. "If I want to sell it, I could do so by signing an agreement in private." She rented out its ground floor as a front shop. Due to the impasse of the resettlement issue, all houses in the village remained intact, including hers. So, she opened a shop by the village back entrance, selling Miao handicraft antiques collected from Miao communities along with factory-made souvenirs. Her shop was the only one that remained open after the PATT closed down its tour there. Looking back, she felt that she had made the right decision to accept the PATT's initial compensation offer. When I mentioned the local Miao village tours, she was dismissive, saying, "The county government abandoned our village which had the most historic value, and tricked tourists to go to so-called 'Miao village' for their money." Those villagers who had refused to move out felt that the closure of their village as a tour site was a punishment for their resistance. They complained that the government ruined their reputation by calling them "*diao min*" (barbaric peasants). "*Guan he min dou, min dou bu guo guan*" (if the officials fight with the commoners, the commoners could never win), one villager expressed their helplessness. According to the deputy director of the county Tourism Bureau, the closure was "due to the fact that the resettlement was implemented poorly." I then asked: why the closure now, as the tour business there had been running all these years with the PATT and the remaining villagers selling their own tickets, despite the conflict surrounding the resettlement issue. She responded affirmatively, saying, "the villagers selling their own tickets, that's what we must *daji* (crack down)." With the PATT's withdrawal, the village lost a high volume of tourists. However, the villagers could have expected fair numbers of tourists coming on their own to visit this well-known historic site without the PATT's involvement. Predicting this, the PATT and the county government tried to prevent any tourists going there. The county Tourism Bureau literally forbade any travel agencies taking tourists there or even recommending it. The deputy director said, "Even though we could not control the villagers, we could control the travel agencies. They have deposit money at both the county and prefecture level!"

At the village entrance, both of the ticket booths by the PATT and the villagers were closed. There was no sight of any PATT employees, and hardly any tourists. The snacks and souvenir vendors disappeared as well. The rampart appeared to suffer from a lack of care, with grass and bushes growing wild on it. The cast iron door put up by the PATT as a ticket checkpoint onto the top of the rampart was chained closed with a big lock. The villagers constructed a simple ladder with cement blocks for the access to its top. They occasionally showed one or two curious tourists up onto the rampart, while the majority of the few tourists who made it on their own to the village would turn around after a quick look at the rundown appearance.

One afternoon around four o'clock, I joined two village women, sitting and chatting by the curb of a house right across from the village entrance. Suddenly, there came noises from outside of the entrance. One of them stood up. "Excuse me," she said to me and hurried out. The other followed her. I went behind them, curiously. I saw a crowd of women was gathering in front of the ticket booth. One woman was in the middle of the crowd, with a stack of cash in small bills in her hands. After she counted the number of those gathered around her, she started giving away the money to them. Whoever came to work anytime on a particular day would receive her share of the ticket sales for that day. With few tourists, they often closed the ticket booth earlier.

Four years ago during my visit in 2006, the villagers were optimistic about the future, and confident that the PATT could not beat them, since "they were just a bunch of outsiders." It might appear that four years later the villagers indeed won over the PATT as the PATT had to close this tour site. Or, it seemed a situation that both parties lost, because the villagers were deprived of the opportunity for their independent tourism development, and it cost the PATT ticket sales from this site. However, the ultimate winner seemed to be the PATT. The PATT offset the loss by turning in less money to the county government than it was required on the lease. It obtained five more attractions to its original eight sites, including a Miao village tour in Longcun.

The Most "Popular"

If you can't beat them, join them. Despite claiming that the current village tour business was having a negative impact on local communities, the PATT was attracted by the increasing market share of *"xiangcun you,"* and obtained its own Miao village tour. The newly added site featured Longcun village. For a full tour experience, the tour guides led the tourists on a hike through the valley, to sightseeing the mountains, caves, waterfalls, and boating through a reservoir before arriving at the village. Ru's husband, the official from the county Water Conservancy Bureau, once commented, "I have never been to Longcun after it became a tour site, but I was once there in charge of water. There was a reservoir, and I remember a tiny road and a cave. Not very good scenery. But I heard that there are a lot of tourists there now. Very good business!"

Longcun is located in Shanjiang Town, north of Tuo River Town, a moun-tainous region with a dense Miao population. As one of the poorer villages in Fenghuang, it is more remote and less accessible than Wucun. It had no elec-tricity until 1996. Nor did it have a motor road prior to 2002. Longcun villagers had relied mostly on feet, taking over an hour, to get to the closest market town of Shanjiang. Even today, no public transportation connects the village to the town. Due to the relative isolation, not until the early 2000s, some villagers started to migrate to the cities on the east coast as wage laborers. Longcun had a total of 176 households. Partrilineal and partrilocal, the villagers (except the in-marrying wives) generally belong to one clan sharing the family name. The pre-migration and pre-tourism Longcun did not contain significant internal dif-ferences in wealth. Its economy was based primarily on subsistence and market agriculture to secure their livelihood. With the rugged and dry *"leigong tian"* (literally "thunder field," a typical karst landform), they depended mainly on rainfall to water the land of rice, corn, and vegetables. Tobacco was once the main cash crop produced for the market, until the Fenghuang Tobacco Factory went bankrupt in 1997. Meat consumption was limited and was provided by raising a few fowl and a couple of pigs. Predominantly poor, the average household ended up with an annual cash income of one or two thousand yuan. Patterns of cooperation, reciprocity, and solidarity were still practiced and were woven into the fabric of the villagers' daily lives. Favors were given and returned often in the form of labor for farmwork, house construction, pig slaughtering for New Years, funerals, and weddings.[5]

In 2007 an outside developer Lu signed a contract with Longcun and obtained the exclusive rights of its tourism development for forty years. Since then, Lu invested to develop and promote Longcun as a tourism village featuring ethnic Miao culture. His major investments included constructing tourist facilities and organizing a Miao performance show. Lu recruited people from around Fenghuang for his village tour business. Hai, who had directed Wucun's performance troupe, moved to Longcun for a month to help Lu put the show together. In 2008, the PATT collaborated with Lu. The PATT bundled Longcun together with its original sites, and sold the ticket for a little bit more to make it a good deal. According to Ma, Lu received an annual fee of 3 million from the PATT in return. Since then, Longcun had been one of the leading village tours in Fenghuang. Longcun's tourists increased dramatically, with hundreds and sometimes a couple of thousand tourists per day during peak seasons (including summer and two weeklong holidays of Labor Day and National Day). Less accessible than Wucun, almost all tourists traveled to Longcun with tour operators' vehicles.

After riding on a tour bus and hiking through the valley, tourists had at most about half a day to spend in the village. Their tour experience was similar to the one produced at Wucun. Their activities in the village included watching a show and having a meal. Depending on their arrival time, they might skip the show, hurry through the village, and head back to Tuo River Town or to another tour site right after the meal. There was little interaction between the villagers and the tourists. After being welcomed by a group of young girls and boys in reformed Miao costumes with rice liquor and songs at the village gate, the tourists gathered around the circular performance ground, the central tourism space in the village, to watch a show. The show was similar to the one performed in Wucun except that it was about half as short. The shortened time was tailored to the tour agencies' needs to conform to an even tighter schedule.

Lu hired around 20 villagers as boatmen, show performers, sweepers, and parking lot attendants, with a salary ranging from 200 to 700 yuan per month. At the price of 8,000 yuan paid to the village for the first year with an increase of 200 yuan every following year, Li monopolized the profit from selling entrance tickets to tourists, leaving a slim margin for the villagers to make money through selling souvenirs or hosting tour groups for meals. The vendors were mostly village women, whose husbands or sons migrated to Hangzhou in Zhejiang working as wage laborers. They complained that although business was good for Lu, it was not so good for them. With hundreds or even thousands of tourists going through the village daily during peak season, few were allowed the time to shop at their souvenir stands. One of them said, "Some of the tour guides are not nice. They told the tourists that the stuff we sell here was more expensive than those sold in Tuo River Town. Some even told the tourists that our stuff was all fake!" When the vendors asked the tour guides to allow tourists to wander freely a little longer, the replies were simply "no time." "No time" seemed to be a decent excuse. According to Jun, the guides could receive high kickbacks from the big shops in Tuo River Town if they took tourists there to shop. Most of those shops were owned by itinerant merchants from outside Fenghuang. With the minimal profit, the villager vendors could barely afford offering any kickback to win over the tour guides and therefore the tourists.

It was unfair to just blame the tour guides for the villagers' diminishing benefit, as their income largely depended on the kickbacks. They were paid 30–50 yuan for each daylong tour assigned by the tour agencies they worked for, either without a base salary or a salary of a couple hundred yuan. The amount they could make from the kickbacks was determined by their individual ability to make tourists shop. Sometimes, the guides had to share the kickbacks with their employers. There was an increasing number of the local youth serving as tour guides. The unregulated yet large local tour market also attracted tour guides from outside Fenghuang. The tour agencies in Fenghuang had little problem to absorb them,[6] as no salary needed to be committed, and more profit could be gained by sharing the kickbacks made by their guides. Agreeing with other private tourism developers in Fenghuang, Lu described the local tour market as "rotten," and said, "in Fenghuang, anyone with two legs could be a tour guide!" Most of the tour guides were indeed not licensed, but they were not the main reason for the problems of Fenghuang's tourism market.

Due to the remoteness and the lack of public transportation, the overwhelming majority of tourists traveled to Longcun as tour groups (Photo 2.1). They stayed only for part of the day. After having one meal and watching the show, they barely had time to browse the souvenir vendors in the central section of the village, or rest for a moment chatting with the villagers. They had to stay in their tour group, rather than dispersing into individual homes,

Photo 2.1.   A tour guide, with a microphone wrapped around his head and a flag in his hand, was leading his tour group going through the tour route in Longcun village. (Photo by Xianghong Feng, 2014).

due to the fear of missing the tour shuttle. It was similar to the mass tour experience production in Wucun, and contrasted markedly with the few tourists
who arrived on their own, lodging and eating with local families in the kind
of intimate interaction that many cultural or ethnic tourists sought. Through
the independent tourists, the local villagers perceived themselves, and were
far more likely to be treated, as active agents/hosts, rather than passive subjects. Just like the villagers in Taquile in Peru (Zorn and Farthing 2007: 681),
they were outmaneuvered by the outside private tourism developer and tour
agencies who had "deeper pockets," access to the Internet, and control of the
transportation, the principal threat to community control of tourism.

Many tourists in Fenghuang disliked their village tour experience. During one
of my visits at Jun's travel agency, a tourist came in to buy a train ticket to return
home. Jun asked him where he went that day. "A Miao village," he said. "Child
beggars everywhere. It felt like a scam. I would never go for another Miao village tour again." However, even though some of them initially decided to travel
on their own, they might change their mind after weighing the limited options
available on-site. Three young tourists came in later asking for information on
a one-day Miao village tour. Shi introduced to them a one-day tour package
including a Miao village and the Heaven Dragon Valley. The price was 108
yuan per person, covering transportation, entrance tickets, and a meal. As they
were a group of eight, Shi offered them a minivan and a private tour guide. They
insisted on only purchasing their entrance tickets instead of a packaged tour, as it
would be "less fun." Shi tried to persuade them, saying, "In that case, you would
have to buy the ticket on site at a price of 148 yuan for each person, plus 200
yuan for renting a mini-van for your eight people. It is a bad deal."

The price difference was indeed significant between traveling on their own
and purchasing a tour package. They did not make up their mind at that time,
and seemed not so sure about traveling on their own anymore. This was a
typical scenario. I was surprised at how cheap the price (108 yuan) Shi offered. Shi explained, "We have to offer the cheapest we can. The competition
is fierce. The tourists will ask around, and then choose the cheapest. If we
did not offer the lowest price, we may very likely lose potential costumers.
But we do need to guarantee about 10–15 yuan profit per tourist to run the
business." Since the space to make money from each tourist's package fee
was being increasingly squeezed, the travel agencies maximized their profits
through making tourism a high-volume business.

IRONY IN "MCDONALDIZED" TOURS

In overscheduled and itinerized mass tours, the tourists are being objectified
as they are hurried through a tour route set by tour agencies and operators, in

which maximizing profit, rather than the local communities' or the tourists' interests, is the priority. It is an increasingly dehumanizing assembly-line production of the tour experience, highly controlled by the tour operators, and along the line the tourists became the "guests" of the tour operators whose value is determined by economic gain the tour operators could expect. Characterized by rigid itineraries, packaged mass tours, as Wang (2006: 68) points out, "embody the rationalism in capitalist commoditization (Weber 1978), and hence act as a rational way in which the tourism industry controls and manages mobile experiences (Ritzer and Liska 1997)."

This is ironic. One major motivation for traveling is supposed to be to escape the "iron cage" (Ritzer 2010b) of everyday life at home. With the "McDonaldized" tour experience, the tourists escaped one "iron cage," only to inevitably fall into another. What is supposed to be leisure time enjoying the unique or extraordinary turns out to be the opposite. Thus, the rationalization of tour experiences defies, paradoxically, the very essence of the tourist pursuit—to escape the overrationalized modern life, and "tourism, the very act of escape from daily constraints, ends up as an alternative constraint" (Wang 2006: 69).

In the current unregulated local tourism market, decent profit for the tour operators and developers could only be realized through replacing independent visitors with a high volume of tour groups. When the economic goals of mass tourism are reached, individual identities for the "guests" became obscured with the label of "*you ke*" (tourists). While guests become dehumanized objects that are tolerated for economic gain, tourists have little alternative but to look upon the toured too as objects with impersonal attitude (Smith 1989: 9–10). In Fenghuang's village tours, the hierarchical relationship of class and ethnicity (Oakes 1995b; Schein 1997) is displayed in the daily scene of the metropolitan Han urbanites gazing quickly at the poor and "backward" Miao villagers as they were hurrying through the village. The interactions between the toured, the tourists, and the tour operators appeared to be increasingly stressful, proportionate to the larger numbers, as it was shown through the quarrels between the souvenir-selling villagers and the tour guides in Longcun, through the frequent complaints from the tourists to the tour developer in Wucun, and through the conflicts between the PATT and Huangcun village.

The tourists' quest for the exotic ethnic culture from Miao village tours was responded to by the tour developers' and operators' manipulation of signs and experiences for their consumers. Therefore, the toured were being objectified and, sometimes, distorted. Miao culture is one of the major local attractions. Non-locals and locals, Miao or not Miao, have engaged in hosting visitors in the name of the Miao to cater to tourism preferences. Tourism could promote cultural exchange, but it could also create misunderstanding. As Bodley (2008: 163) points out in his discussion of tourism and indigenous peoples, "tourism creates and maintains illusions about exotic cultures." The tour mediators expedited culture change with the misrepresented Miao and other local cultures in the inaccurate marketing brochures, misinforming displays, staged shows, and tour guides' scripted stories and jokes about the Miao to entertain tourists. Visitors in their "over-packaged" tours were presented with a "pseudo-cultural" representation of the Miao.

Although often advocated as an antidote for poverty (WTO 2006), tourism in the impoverished region may be a product of national economic inequality and often disproportionately favors a few special-interest groups (Bodley 2008: 162). In the case of Fenghuang, the few special-interest groups, including the PATT, other tour developers and operators, and local authorities, formed elite alliance and business networks, and they were the actual "hosts" of Fenghuang's tourism. Wucun tour (controlled by a local tour operator) and Longcun tour (monopolized by a non-local developer with the collaboration of the PATT) largely excluded the majority of the local villagers; Huangcun, a significant historic attraction, became an abandoned tour site due to the villagers' fight for the right to host their village tourism. With the growth of village tours propelled by the elites, ethnic minorities such as the Miao are further marginalized by tourism that is supposed to do the contrary. With packaged mass tourism being selected as the development path, through a top-down decision-making process typical in tourism planning in China (Wang et al. 2010), it contradicts the original official purpose of developing "*xiangcun you*"—the improvement of the living conditions of rural residents.

EFFECTS OF NUMBERS

In a neoliberal market economy, profit is placed over people (Chomsky 1999), and "command over spaces and times is a crucial element in any search for profit" (Harvey 1990: 226). Tourism industry is no exception. With their effort to McDonaldize mass tourism, tour operators fabricate tour packages featuring highly standardized and overscheduled itineraries, which realize their control of touristic space and time, and therefore, maximize their profit. As demonstrated in the village tours discussed above, it is an increasingly dehumanizing assembly-line production of the tour experience, strictly controlled by the tour operators in terms of length of stay, choice of lodging, what to see, where to eat and shop, and who to interact with. The increased volume of tourists and reduced profit per tourist is likely to exacerbate the existing leakage of profits away from the community, and provides a diminished experience since the toured and tourists have much less interaction.

Smith (1989:11–16) categorizes seven types of tourists from explorer to charter tourists to discuss "the effects of numbers," and states, "Tourists themselves can become a physical as well as social burden, especially as their numbers increase." The increased scale of mass tourism added layers of mediation to the fading host/guest relationship (Chambers 1997: 6), in which private tourism operators, motivated by profit, tend to push larger and larger volumes of tourists through destinations with little regard of the place, people, and culture they are consuming. If the nameless and faceless mass tourists in loaded coming and going tour shuttles were blamed for bringing physical and social burdens on their destinations, the fundamental problems underlying the "McDonaldization" process in mass tourism might be obscured.

Bodley's (2003) power and scale theory suggests that scale gives elites a powerful incentive for growth, as increase in scale (e.g., the size of population, markets, and economic enterprises) tends to disproportionately benefit those at the top of any social hierarchy and socialize the costs among the majority at the bottom. Acknowledging the "power of scale" helps to better understand the "irrationality" innate in the overrationalized "McDonalization" process directed by the elites, in which people's creativity and freedom are constrained or eliminated (Ritzer 2010a, 2010b), and all types of social power are concentrated in the hands of a few (Bodley 2003). In the global context of "scaling up" being adopted by the policy-makers as an ideal development paradigm to alleviate poverty (Sachs 2005), such analysis here with contradicting results is particular relevant, and it has broader implications in sustainable development beyond tourism.

NOTES

1. This chapter is revised from "Who Are the 'Hosts'?: Village Tours in Fenghuang, China," which appeared in *Human Organization* 71(4): 383–394. Reproduced by permission of the Society for Applied Anthropology from Human Organization.

2. It was an adaptation of the Miao custom for receiving guests.

3. The money used for house renovation or rebuilding was earned and saved through the years by the villagers who worked as migrant laborers in cities, not from Wucun's tourism development. Comparing with many other Miao villages, Wucun had a relatively easier access to the market town, and the villagers started to migrate to cities for wage jobs in the early 1990s.

4. Huangcun village was first built in AD 687 during the Tang Dynasty. The stone rampart around the village was added around the 18th century during the Qing Dynasty and has been well restored. The rampart is 686 meters long, 5.6 meters tall, and 2.9 meters wide, with three gates at the direction of the north, east, and west.

5. For a general discussion of various forms of reciprocity among the Miao communities in Fenghuang, see Fenghuang County Ethnic Groups Gazetteer Writing Committee (1997: 77–81) and Shi (2002: 101–105).

6. In the late 1990s and early 2000s, there were literally no travel agencies in Fenghuang. By 2011, there were seven travel agencies and twenty-four travel agencies' service branches.

Part II

LIVING WITH TOURISM

Chapter Three

Spatial Transformations

Constructing Tourism Sites

Mu, an old Miao woman vendor, was napping at a corner in the *Gu Cheng* on a hot summer day. Surrounding her were factory-made souvenirs on display in her basket and on the nearby ground. Business for her had not been good. She missed the old days when she sat by the Rainbow Bridge, making and selling the Miao-style paper-cut handicraft, formerly a signature scene in the *Gu Cheng*. With Fenghuang's tourism boom following 2002, the county government had forbidden her doing business by the bridge, but reluctantly allowed her to move to this corner. Her handmade paper-cut pieces became harder and harder to sell when cheap factory-made souvenirs saturated the local tourism market in migrant merchants' shops that flourished in the most profitable space frequently visited by tourists.

Mu was my key informant on traditional Miao paper-cut handicrafts during my first visit to Fenghuang in 2002. Sitting by the Rainbow Bridge doing paper-cuts every day, she got to be well known by the local residents and occasional tourists. Displayed by her side were exclusively paper-cut handicrafts in various designs. Dressed in Miao-style clothes with her headband and diligently cutting designs on paper, she and her stand became a favorite subject of the camera lens, and was even featured on one of Hunan's provincial television shows. The large-scale tourism development was yet to take shape in Fenghuang. Including Mu, few had predicted the dramatic change that was just around the corner for this place and its people.

With the YDCC taking over the management of Fenghuang's main tourism attractions, masses of tourists flooded in. The original shops along both sides of the historic cobbled streets selling local items were taken over by itinerant merchants who renovated the shop interiors and replaced the local items with factory-manufactured goods from around China. Amid all these changes, I lost contact with Mu, and eight years had passed by. One summer day in 2010, I was on my way to the *Gu Cheng* looking for a quick bite to eat. A vendor's stand caught my attention. It was in a street corner by the entrance of an Internet bar. Two umbrellas hung together with a blanket created a shady spot under the hot midday sun. Smaller goods such as water guns, shuttlecocks, machine-embroidered wallets and cushion covers were displayed on the ground and in small baskets for sale, just like many other souvenir stands nowadays in Feng-huang. I would have walked by if I had not noticed the few paper-cut pieces. It was her, Mu, the vendor who retreated into the shade behind her stand.

Mu was no longer allowed to place her stand by the Rainbow Bridge, as the county government started to implement spatial regulations that restricted where local street vendors could operate. The government designated specific locations to the vendors for central control, and required them to pay a monthly fee for a fixed spot in the officially designated space. Mu protested to the county government, and argued that she should receive a vending space free of charge because of her contribution to promote local tourism as one of the well-known Miao handicraft artisans. The government then allowed her to temporarily place her stand at this corner. After all, the official Tourism Guide magazine by the county Tourism Bureau was still featuring her and her paper-cut handicraft. The irony was that Mu was no longer engaging in paper-cutting. She told me that there were more and more tourists in town, but her paper-cut business was worse and worse. Mu stocked her stand with machine-made cheap souvenirs bought from the local wholesale stores. When we had a photo taken together, she took her paper-cut tools out and posed as if she were still practicing it. This chapter expands the discussion in Chapter 2 on the temporal-spatial control of tourists by tourism mediators, to examine how power and social transformation are embedded in the tourism space.

SPACE, POWER, AND TOURISM

Studies on space have proliferated in the social sciences, and the interconnections between social relations and spatial structures have become a central focus for scientific inquiry especially in sociology and geography (Gregory and Urry 1985). Much of the earlier research concerning the relationship between space and power has focused primarily on theoretical development with concrete empirical details missing or being dealt with at an abstract level, as pointed out by Lawrence and Low (1990). In spite of the importance of spatial process in local politics and social change, the productive role of space and its intrinsic link to power dynamics were not often explored in anthropological analysis until the early 1990s (e.g., Moore 1996; Rofel 1997; Zhang 2001a, 2001b; Yang 2004).[1]

Despite the increasing anthropological discussion on space as part of the recent "spatial turn" in social sciences (Jimenez 2003), tourism space has not hitherto been primary in the theoretical concerns of power relations. Space and power are not always analyzed in the context of tourism from an anthropological perspective. The existing scholarship on power and tourism space largely reflects a geographic approach (e.g., Church and Coles 2007; Edensor and Holloway 2008; Gatrell and Collins-Kreiner 2006; Teo and Leong 2006).[2] These studies can be criticized for a lack of thick ethnography and tendency to focus on the spatial experiences of the visitors/tourists, setting aside the consideration of the tourism brokers and local residents.

The relationship between space and power in tourism analysis deserves further conceptualization and contextualization. In this chapter, I explore the contests and tensions between various actors and interests in the construction of tourism space. I mainly draw upon Foucault (1975, 1978, 1979) and Lefebvre (1991) to show how local tourism spaces became dominated by the public/private brokers at the expense of the local villagers who experienced varying degrees of exclusivity from the profitable tourism resources, especially the access to tourists. As Cartier (2002: 89) states, while both laying stress on human agency, Foucault is especially concerned with conceptualizing power relations in situated contexts, and Lefebvre focuses more on the relationship between the state and private capital interests and how this alliance creates the urban-built environment.

Foucault's work is utilized here in two aspects. First, I echo Cheong and Miller's (2000) conception of applying Foucauldian power relations (1975, 1978, 1979) in tourism analysis, in which the tourists are considered as "target" (much as Foucault portrays the incarcerated criminal in prison) rather than "agent" in tourist destinations. I divert the attention from the tourists to instead focus on the agency of the public/private brokers and local villagers, emphasizing their distinct spatial practices. I recognize that the brokers are a dominant force in the control of tourism development and tourist conduct in the local reality, but this is not to deny the villagers' contribution to the transformation of local tourism space, as they constantly negotiate and contest the direction of development in the pursuit of their rights and interests. Secondly, Foucault's (1979) concept of power as being created through architectural structures (e.g., the panopticon) that shape subjectivities is useful in analyzing the control of tourism space through the design and construction of built forms.

Lefebvre's (1991) ideas on production of space are instrumental to unraveling socio-spatial dynamics in a tourism context. Lefebvre (1991: 33) outlined three dimensions of space as spatial practice (the perceived, referring to policies, activities, rules, and structures), representations of space (the conceived, referring to space or conceived geographies), and representational spaces (the lived, referring to place or lived geographies). This triad, as pointed out by Gatrell and Collins-Kreiner (2006: 767), recognizes the equality of space, place, and practices, in which the decisions, policies, and activities (spatial practices) of different actors in space transform conceived geographies (representations of space) into lived experiences (representational spaces). It realizes a unity between physical (*real*, as concrete reality of space), mental (*imagined*, as abstract thought of space), and social (*real-and-imagined*) space (Elden 2004: 190).

To illustrate the mutual constitution of space and power relations, I apply Lefebvre's (1991) triad to analyze the spatiality of everyday life in Fenghuang, and draw attention to the distinct spatial practices and representations of spaces between the two main actors—the tourism brokers and the villagers, who have both shared (in the sense that the villagers had to rely on the brokers' deep pockets to bring tourists in) and conflicting (in the sense that the villagers and brokers compete for access to on-site tourists) interests in economic gains from the tourists. It is important to understand that representational spaces reflect spatial practices and representations of spaces for both the brokers and villagers, rather than from the sole will of the brokers who occupy a higher status in the local social hierarchy.

POLITICAL REGULATION AND
PLANNING IN THE *GU CHENG*

As discussed earlier, upon leasing out tourism attractions to private developers, the county government appropriated public resources (e.g., mountains, rivers, and historic sites) that had been shared by the local residents and transformed them to private properties monopolized by the developers exclusively for profit. Access to tourists became a central focus of competing claims between the developers and local peasants. The county government imposed a series of spatial zoning regulations to enforce the privatization, intending to keep the peasants out of the touristic space and hence to limit their access to tourists. With the implementation of these zoning plans, new concepts such as "wild tour guide" (*ye daoyou*), "wild boatman" (*ye chuanjia*), and "illegal solicitors" (*feifa lakede*) were created through official discourse to label the local peasants who continued to practice their informal tourism businesses, which consequently became "illicit" activities.

Scott (1989: 9) applied Foucault's (1979: 274) concept of state-created "crime" to discuss peasant resistance in the form of poaching, an activity that had been part of the traditional subsistence routine of the rural population and that was embedded in customary rights. Such analysis is applicable here. The "illicit" activities mentioned above reflect less a change of behavior on the part of the peasants than a shift in local property relations and power structure. Under pressure from the developers, it is the local government and its creation of new zoning regulations that suddenly transformed the local peasants' practices of informal tourism businesses into "illicit" activities, thus, becoming everyday forms of resistance.

The reluctance of the local authorities to use coercion in strictly implementing the spatial zoning left room for peasant resistance to persist. Besides, there was a certain sense of popular justice that "you have to let the poor make a living." Harsh means (e.g., raising penalties, prosecuting noncompliance, and appointing more enforcement personnel) were not ideal solutions for the local authorities to eradicate these newly defined "illicit" activities. The county officials were concerned that high-handed policy enforcement would incite open protest and intensify conflict, contradicting the image of a "harmonious society" (*hexie shehui*). As higher-level officials frowned upon such incidents, the county officials lost major points on their end-of-term political performance evaluations.

Stopping the "Chaotic"

In an earlier publication (Feng 2007b: 22), I discussed the competition between local street vendors and non-local shop owners, which intensified into direct conflict when the shop owners pressured the government to forbid the "illegal" mobile vendors from doing business in the street, as they were indeed not licensed. In the name of restoring the spatial order in the overcrowded *Gu Cheng*, the county government forbade the mobile street vendors through the force of *Chengguan* (the City Administration and Law Enforcement Bureau). In order to implement the spatial order of the marketplace, the *Chengguan* personnel increased, so did the daily quarrels between them and the vendors. One *Chengguan* patroller complained that the job was tough physically and mentally, "Three shifts everyday with five and a half hours per shift, plus getting scolded by people everywhere!" He was having a short lunch break in a small restaurant in his work duty zone in the *Gu Cheng*. "We have to find lunch ourselves. Our *danwei* (work unit) doesn't make arrangements to provide us meals, as we are too dispersed." After quickly finishing his bowl of rice noodles, he rushed back to patrolling the street.

However, increased patrol by the *Chengguan* seemed to be a fruitless effort as it was almost impossible to prevent them from reappearing in the street again, only resulting in more daily disputes between the government and local residents. Therefore, the county government designated two areas for the mobile stands: along the sidewalk by the Tuo River, and a part of the street toward the Rainbow Bridge, both of which were away from shops on the main street. Spatially disciplined, long lines of baskets arranged tightly side by side with each other displaying small souvenirs. Each basket was a stand, with the vendor sitting on a little stool behind it. An official from the county Government Office explained, "The vendors cannot just set up their stands as they please. It is bad for the tourism market order, and bad for the environment. They have to obtain a temporary permit and to be regulated at a fixed spot. In the past, they were all over the streets, totally chaotic!" The large number of stands confined within limited space had resulted in maybe more "chaos" as they literally crammed the narrow street and sidewalk, making it hard for pedestrians to pass through, let alone shop.

It did not take long before the vendors were relocated. By the summer of 2011, almost all the vendors disappeared from the main streets in the *Gu Cheng*. The county government had built a line of wooden stalls along one of the side streets for the local mobile vendors to rent. Many of the stalls were closed and locked. At the few that were open, the vendors set up baskets to display their goods outside their stalls, a desperate effort to attract tourists' attention. One of the vendors told me that she was convinced when the government announced that no other places would be allowed to have stands after last October, and all vendors would have to do their business here. "The government made it sound very nice, saying that the tourism bureau will require the tour guides to take tourists through here. Now almost a year has passed by, not many tourists at all, only occasionally random ones. The majority of tourists were taken by the tour guides to visit the big shops in the central streets," she complained. "We vendors have to rent the stall for at least three months each time at a rate of 200–300 yuan per month. I am here for over a month now. Once the three months is over, I am not going to renew it!"

Political spatial ordering of the local street vendors resulted in a decline in the number of these mobile vendors. Spatial fixity facilitated and reinforced the political control, and decreased the flexibility innate to informal tourism sectors in which the local ordinary people, possessing little financial and social capital, were able to participate. The additional monthly fee further squeezed their already marginal profit. One of them gave me an example: an engraved copper paper weight used to be sold at 70–80 yuan by the local vendors, but later, even as the price dropped to as low as 15 yuan, it was hard to sell. It cost the vendors 9 yuan to purchase it from the wholesale shops, and they had to pay the government a monthly fee. "It is very hard to do a small business like this. I have done it before, so I know. The tour guides tell their tour groups that the vendors' goods are all fake, and then take them to the shops of the non-locals. But the stuff there is all fake too." She was a street vendor before working in a noodle restaurant to clean tables and wash dishes. Many local souvenir sellers like her switched from their small-scale entrepreneurial endeavors as mobile vendors to other insecure temporary wage jobs.

Image Projects

Since the early 1990s, local government leaders in China have competed with one another to build public projects such as new city halls, central plazas, and statues commemorating local legends, to visibly demonstrate their achievements (Hsing 2008: 60). Just like many other local officials in China, the ultimate concern of the Fenghuang government leaders was the political performance evaluation at the end of their term. The manifestation of political performance occurs through the tourism landscapes created by investment capital, and the ability to attract investment capital has become the ultimate symbol of political power (Lew 2007). Image projects (*mianzi gongcheng*), the symbolic projects in strategic locations to improve the physical appearance and built environment of the place, were prioritized over the local people's lives tied to this place, since the former could turn into political accomplishment easier and faster as they were more visual, direct, and seemingly more permanent.

Ru's husband introduced me to the county's "ten-mile sightseeing belt" along the Tuo River, which he was in charge of as a high-ranking official at the county Water Conservancy Bureau. Beginning in 2006, the county government started investing in this project to increase sightseeing spots along the river, expanding from the central part of the river surrounding the historic North Gate monopolized by the PATT. By 2010, 4,000 meters of wooden railings were added along the riverside, and one big and several small wooden bridges across the river were built as tourist attractions and also for tourists' convenience. Noticeably, new buildings emerged along the upper riverside stretching a couple miles from the North Gate. As soon as the "belt" was starting to be built, it immediately increased the business value of the land along it. Residential houses were replaced by commercial properties, especially tourist hotels and bars.

I asked Ru's husband, "How is the business of the new tourist hotels?"

"A lot of business, and the occupancy rate is very high. There are so many tour groups, and each tourist hotel connects with individual tour operator specializing in packaged tours," Ru's husband replied. He continued, "We just need make sure the work is done right for the flood control."

Ru's son worked at the county hydropower station, the same system with his father's bureau. Ru's daughter-in-law was a hairdresser at a hair and beauty salon. After the sightseeing belt was built, Ru's son and his five friends spent 2.4 million yuan and bought a bar at the second floor of a riverside property. They renovated it and renamed it "Casual Encounter." Ru's son invested 300,000 yuan. He and his wife came to the bar after their regular daytime job, and worked until two to three o'clock in the morning. Ru's daughter Yu once took me to the bar when she was in town visiting her parents. I had known Yu since she graduated from high school. Over the years, she finished college, found a job, and moved to Changsha where she got married and just had her first baby. Not many guests were at the bar that night. Yu said her brother hoped the business would gradually pick up. According to Yu, it was expensive to keep the bar open, as the annual rent was close to 100,000 yuan, and the monthly salary for its lounge singer was 9,000 yuan. "It cost too much to run a business in Tuo River Town nowadays," Yu said. One friend of Yu was interested in a front shop, but couldn't afford the price of over one million. A year later in 2014, the flood in Fenghuang brought severe damage to most of the riverside businesses, including this bar.

Another major "image project" was the Culture Square (*wenhua guangchang*) in the *Gu Cheng*. It was a newly constructed plaza with a phoenix sculpture in the center. It stood as a symbolic mark for a phase of rapid urbanization in Fenghuang. It soon became one of tourists' favorite sites for souvenir photos, where local camerawomen gathered for their ethnic costume rental and photograph business. As the local tourism market has grown increasingly dominated by outside developers and merchants,[3] souvenir selling and renting ethnic costumes to tourists for photographs were among the few informal tourism sector activities left to the local ordinary and poor. With the government putting more regulations on the vendors, the costume renting and photograph business was seemingly the only "prosperous" one left.

Unlike many souvenir vendors who need a spot to display their goods, the local camerawomen were more mobile. They placed their baskets with rental costumes in nearby photo-printing shops, with which they normally developed a business relationship (see a detailed discussion on their cooperative ties in the next chapter). Then, they only needed to hang a camera on their neck and hold a plastic laminated brochure in hand, on which were displayed sample pictures of tourists dressed in ethnic clothing posing in front of local scenes. They gathered at the must-see sites, soliciting tourists to rent ethnic clothes and have their photos taken by their professional-looking digital cameras. Five yuan or less for renting clothes, and ten yuan for each picture tourists decided to purchase. On a good day, it was possible to make a couple hundred yuan if one or two of their customer(s) happened to like more than a few of the photos and were generous enough to purchase all the ones they liked.

The decent profit again attracted attention from private developers. Liang was concerned that she might not be able to continue to be a camerawoman, as she heard that someone was about to monopolize this ethnic costume rental and photograph business. Indeed, the county government put for bid the exclusive rights of this business at popular photo sites in the *Gu Cheng*, including the Culture Square. A company named Fenghang Shipeng Photography Co., Ltd (FSPC) won the bid. The FSPC then asked the local camerawomen to register, despite their resistance. In March 2013, when the *Chengguan* tried to kick these camerawomen out of the Culture Square, they united together, and "a bunch of women drove away the group of gangsters (*xiao hunhun*)," as Liang's husband, Hong, described. Hong thought that it was almost impossible to implement the monopoly of this business, as there were hundreds of local camerawomen like Liang who we met in Chapter 1. "The government couldn't forbid them by force, as they did not break the law after all," Hong said. By the summer of 2014, Liang and her fellow camerawomen were still doing their businesses.

TRANSFORMING LONGCUN[4]

Schein (1997: 84) points out the importance of recognizing the heterogeneity among Chinese minority communities, and she states "status distinctions penetrated far into the rural hinterland and affected the hierarchical ranking even of remote villages." In Longcun, economic changes accelerated by tourism under the control of Lu have made the community less homogenous. Social stratification in the village is now economically based. Important variables in this process include the spatial differentiation of village land (both residential and agricultural) and the political status of the villagers and their kinship ties. Both variables affected the villagers' access to tourism opportunities as a supplement to household earnings.

Since Lu acquired control of Longcun's tourism development in 2007, he wasted no time to implement his tourism plan in the village. Lu's first and foremost project within the village was to construct a performance space for his Miao show troupe to perform for tourists. Lu's initial attempt to purchase a piece of cropland close to the village entrance encountered strong resistance from the villagers who farmed it. In the end, Lu rented land from nine households. If the land was a rice field, the rent equated to the price of its annual production of rice; otherwise, the rent was 150 yuan per household per year. Looking back, Lu complained how difficult it had been to "educate" the villagers, who were characterized by him as very "*sixiang luohou*" (backward-minded).

Lu eventually succeeded by relying on the support of the village cadres, especially the village Communist Party secretary Qiu. Recounting that experience, Qiu said, "When Lu first came to our village, I worked with him on tourism development planning including the performance ground. It was hard, and almost half of the villagers were mad at me. For a long time, they stopped greeting me when they saw me; and when I called their names, they did not want to respond." He laughed and continued, "Some start to see the tourism benefits now, and it doesn't seem that they still remember their grudge towards me anymore." Just like Lu, Qiu attributed the villagers' resistance to the Miao people's lack of *wenhua* (literally means "culture," and here it specifically refers to the Han culture, especially those associated with modernity).

"No Trespassing While the Show Is On"

As the villagers had no share from the entrance ticket sales, which were monopolized by Lu, they had to find their own ways of conducting tourist business to make money. Their involvement included running family restaurants, operating rental vans, and most commonly, selling snacks, drinks, and trinkets. Access to tourists became the key to their overall income. First, Lu obtained village lands to construct the circular performance ground. Shortly thereafter, some villagers entered this tourist-concentrated area, either setting up mobile stands, or directly approaching tourists to sell their goods. In response, Lu built a performance stage with a two-story wooden shelter flanking both sides with seating on the second floor for the tourists to watch the show. The set of new structures created a central tourism space in the village and limited villager's access to tourists (Photo 3.1).

Photo 3.1. The performance area constructed by Lu as the central tourism space in Longcun village. (Photo by Xianghong Feng, 2011).

Rofel (1997: 159) argues that "space—and the authority to construe it—is a contested domain of relations of production because of its recognized connections to power." In this case, the invisible boundaries between Lu and the villagers were being materialized and displayed through the construction of a central tourism space. Space contains power relations in the form of the built environment. Rooted in imperial China, the symbolism of the center was constructed as the nexus of inner/outer and upper/lower, and what was most inner equaled what was most upper (Rofel 1997; Zito 1997). The architectural design exhibited and reinforced Lu's power: the raised stage and seating areas helped secure on one hand his control over the tourists by keeping them in, and on the other hand his discipline over the villagers by leaving them out. Spatial relations render the power to shape subjectivities. Architectural design of the performance space clearly defined the village's tourism center as "private" instead of "communal." It served as a manifestation of hierarchy, facilitating Lu's monopoly over tourists, and thus tourism business profits.

As Lefebvre (1991: 56–57) points out, "abstract space . . . has something of a dialogue about it, in that it implies a tacit agreement, a non-aggression pact, a contract, as it were, of non-violence." Realized by the architectural design, the performance area, as the village tourism center, generated an imposed "consensus" that such a space is supposed to be quiet and "trouble-free" for the tourists to enjoy their time. Even though the possession and consumption of the communal village space could not be entirely privatized to exclude the villagers, Lu succeeded in enforcing a distance between the tourists and the villagers in a "subtle" way. Consequently, there was to be no constant fighting over keeping the villagers out to maintain the "order" desired by Lu, from whom the message was clear, "no trespassing while the show is on," as he once explained to me.

Lu denied there was much conflict between him and the villagers, claiming "the villagers are generally supportive of me, because I brought tourists in and they are benefiting." It was a different story from the villagers' side, and their responses were mixed. A few village elites expressed their strong position in support of Lu, as Jing, the owner of one of the six family restaurants in Longcun, put it, "It is all because of Lu [that we are benefitting]. So we should protect him. Protect him, then we can all stand up; otherwise, we could never be up on our feet."

Some villagers complained about Lu and his way of running the business. Others, however, engaged in different forms of resistance: a villager sneaked onto the performance stage at night, and damaged one of Lu's Miao-style performance drums; another slapped a tour guide's (one of Lu's employees) face during a quarrel; a few blocked the entrance road for tour buses; and some more actively tried to challenge the boundaries laid down by Lu to gain access to tourists. Political pressure on the villagers was constantly extended from the officials at the village and sometimes even the township level. Despite the efforts of local officials to assist Lu in carrying out his tourism plan and to mediate conflict, tensions between the villagers and Lu intensified, centering on the allocation of the newly constructed stalls for village vendors.

"Keep the Village Vendors in Order"

Lu's endeavor to mold the village's tourism space did not stop at the performance center. Claiming that it was his effort to maintain order and improve tourists' experience, Lu launched the construction of about 120 vendor's stands near the performance center and the parking lot for tour buses, where tourists normally spent most of their time. Lu planned to offer these free stands to those villagers selling goods to tourists, under the condition that they must obey his management. He explained to me, "My intention was to keep them in order, and that's all." These stands were under construction during my visit in Longcun in 2011. The new stands were quickly taking shape, so was the tension within the village. Seemingly a wise strategy, Lu left the village tourism leadership team, composed of the main village cadres, to handle the allocation in an attempt to shift the conflict away from him and toward an internal conflict within the village itself.

With this thorny issue in hand, the village tourism leadership team was particularly concerned. Mi, in his late fifties, had been the village treasurer for more than ten years and was a member of the leadership team. He prioritized the stand allocation as the most urgent problem facing the village officials, saying, "This problem is right in front of us. We have to face the reality. We are very worried, so are the township government and Lu. Whether it could be handled well or not is critical. If it goes well, it will be a very good thing; if not, then very bad, and Lu might have to close his business here. We are a big village, and this is directly relating to the benefit of our more than 100 households."

The 120 newly constructed stands were approximately adequate for the villagers who were currently involved in selling small goods to tourists. Lu announced that he had no intention to find space for those who were then, or would be later, interested in this business, "They have to find some space themselves. If they could find a spot, I am willing to build each of them a similar stand with the standard size of 5 square meters for free; if they couldn't, I can't help with that." Mi's remark helped explain the situation, "If you insist on Lu finding a space for you and he does, you are likely not willing to do business there anyway. No one will be interested in a spot where there are few tourists. There isn't that much good space after all. How can the limited space hold more stand units! You have to understand even though you don't."

The villagers seemed to have no choice but to accept Lu's plan, while waiting for the stands to be built. Stand allocation, the potential trigger for the outbreak of an intensified village-wide conflict, was the village officials' big headache. While the completion of stand construction was only days away, the allocation plan was still far from ideal. Based on Lu's intention, those who had already been selling goods to tourists would be guaranteed a stand. Therefore, a few would be left available to the interested villagers who were newcomers. The original plan was to settle it by casting lots, to determine the order of who would have first choice. For those with a guaranteed stand, the order would be the key to pick a better location; for those competing for a stand, the order would make all the difference. Those who had already been doing business on a good spot were against casting lots, and the others were for it. The divided positions aborted the officials' original plan. This village-level issue caught the attention of the upper-level political administration. The town government in Shanjiang sent officials to the village to assist the village tourism leadership team. In consultation with Lu, they worked out an alternative to allocate most of the new stands according to the existing vendors' current spatial order. Each new stand would be numbered, and the individual villager got the stand on or close to the spot where he or she had been doing business before.

Much discussion on the stand allocation was still going on within the village by the end of my stay in 2011. This was undoubtedly a big issue. For the villagers, it was a matter of livelihood. "Even though it is only a stand, for the left-behind older villagers who could not work as migrant laborers, having a stand is the guarantee for earning a little more cash a year, and not having it means no chance for any tourism income at all," as one villager put it. For the village cadres, it was a matter of social stability, as it already caused friction among the villagers and between them and Lu. More importantly, their personal economic gain, derived from their political status, depended on Lu's success. Among the owners of the six family restaurants providing standard meals to tour groups, the single most profitable tourism business not monopolized by Lu, one was Mi (the village treasurer), one was Qiu (the current village Communist Party Secretary), and another was Zhi (the former village Communist Party Secretary). The highest political authority at village level is the Community Party Secretary. Therefore, as the most influential man in the village, Qiu was offered by Lu a managerial-level position in his tourism company located in Shanjiang, with the majority of over 100 employees hired from outside Fenghuang. In the company, Qiu was mainly in charge of solving conflicts, especially those between Lu's company and the villagers.

The stands were never officially allocated. One major issue related to the twenty villagers employed by Lu. Based on the proposal of the village tourism leadership team, these villagers shouldn't be allowed to take a stand to conduct a vending business. "Warn them first, fire them if they continue," Qiu said. "One shouldn't take a job at Lu's company and run a stall at the same time." Qiu continued, "For example, there was a girl who recently married into our village. Lu hired her to conduct the welcome ceremony for arriving tourists at the village entrance. She set up a vending stand by the performance ground. When she saw tourists walking towards her stall, she would run from the entrance to her stall, while still singing welcome songs. She did not put heart into her job, and it gave the tourists bad impression." The town government once called for a village assembly. The assembly was scheduled on one night in April 2012. It was eventually cancelled due to heavy rain. Many villagers thought that the assembly was pointless. They told Qiu, "If those who are employed by Lu accept it, we have no problem whatsoever." Those employed by Lu said to Qiu, "It is neither the government nor the village who gives us paycheck. Our salary was from the company, and the company is fine with it." Qiu then spoke with Lu, requesting him to arrange a meeting among the twenty villagers in his company. "The meeting never happened," Qiu told me. "So we could do nothing."

TEMPORAL-SPATIAL CONTROL
THROUGH TOUR ITINERARIES

In China, both local residents' and tourists' spatial movements and practices could be heavily controlled and managed by political and economic forces (Su and Teo 2008; Svensson 2010). The above discussion on the spatial practice in the *Gu Cheng* and Longcun in Fenghuang demonstrates that altering the ways that space was used and defined served the elites' interest of monopolizing profit. Time is money. Time, interconnected with space, demands equal attention in the process of pursuing profit. In the tourism market, those who can affect the temporal-spatial distribution of the tourists can reap material rewards most effectively. As Harvey (1990: 214) says, "The common-sense notion that 'there is a time and a place for everything' gets carried into a set of prescriptions which replicate the social order by assigning social meanings to spaces and times."

A key feature of packaged mass tours lies in their control of tourists' consumption through standardized schedules to construct and implement tour time and routes. Thus, in a local tourism economy, "tourism space" and "tourist time" are created through the fabricating of tour itineraries. Mass tour itineraries are often tightly ordered as an assembly-line experience that quickly moves tourists through various venues to favor the interests of the tour developers and operators over those of either the tourists or the toured.

Wang (2006) explores itineraries' significance as "temporal-spatial carrier" from the experience of the tourists. Along the same lines, I look at itineraries here as a "temporal-spatial control" from the perspective of the toured. The criteria of inclusion and exclusion of tourism resources (places and people) in the itineraries was determined not so much by cultural factors (as might be expected in culture tourism), but by the economic consideration of the tourism mediators. The local vendors' experiences in both the *Gu Cheng* and Longcun demonstrate that temporal-spatial control through itineraries realized its maximum efficiency when combined with spatial ordering imple-mented through the political and economic power possessed by local officials and private tour developers.

In Longcun, along with the performance of everyday tourism-related ac-tivities, progressive monetization of relations in village life was transforming not only its space, but also time. This process, expressed repeatedly in the routine daily practices related to tourism, announced a new socioeconomic order in the village. The villagers were hampered in the practice of their everyday lives. Their family visitors could not enter without being met and escorted, as the entrance ticket checkpoint set up by Lu restricted all outsid-ers' access to the village, as is also documented from the tourist water town of Wuzhen in eastern China (Svensson 2010: 218–219). Furthermore, tour-ism interrupted the cycles of agricultural and lineage-based production that provided the basis for social interaction in the community, much like to the case of the Hmong village in Thailand discussed by Michaud (1997). Under Lu's control, the profit from the villagers' small-scale tourism entrepreneurial endeavors, such as selling souvenirs, was rather marginal, and therefore, the villagers continued to rely on subsistence activities to ensure household food security. However, tourism involvement imposed considerable constraints on their ability to maintain their levels of farmwork. They had to adjust their daily schedule around the tourists' activities.

The village's tourism day started every morning whenever the first sound of the Chinese shawm (*suo na*) arose from the village gate announcing the arrival of the first tour group of that day. The essential part of the village tour was the show performed on the newly constructed stage. The end of the show was the villagers' cue to quickly catch tourists' attention with whatever they were trying to sell as the tour guides hurried their tour groups through to the next item in the overscheduled itineraries. The villagers' hours of waiting was in trade for maybe a few minutes of business opportunity with the tourists. For the families involved in catering standard meals to the tour groups, the ending note of the show was the signal to have the dining tables served with dishes they had prepared all morning in the kitchen. Loud music through speakers was repeatedly played throughout the day until late afternoon when the last tour group left. Only then did the village fall back to its normality: noise and chaos retreated with the coming of the nightfall; against the restored tranquility were the occasional sounds of children playing and dogs barking; and the smell of firewood smoke arose from the mud-brick houses built along the hill slopes. The irony is that this pastoral scene, which was sought by the urban tourists, painted by the tourism brochures, and promised by the tour developers, always retreated with the tourists' arrival, and only returned after their departure.

SPATIAL PRACTICES

I summarize in Table 3.1 how the decisions, policies, and activities (spatial practices) of the main players in Fenghuang's tourism space (i.e., the tourism brokers including the political authorities and private developers, and the local villagers) transform conceived geographies (representations of space) into lived experience (representational spaces). The local political authorities and outside private developers are characterized by their similar efforts to restrain the bodily movement—mobility—of the local villagers (the tourists as well); the villagers mainly relied on the practice of their "hanging out" (Teo and Leong 2006) to resist such restraint, tactically using their own mobility to their most advantage. Placed within touristic power relations, the juxtaposing of spatial practice and representations of space of the two main actors as presented in Table 3.1 helps unfold the process of an intensification of marginality in representational spaces, in which the powerful enforced their hegemony and affected the lived experience of the weak, despite their resistance.

The county government's regulation of tourism space (e.g., spatial zoning and *Chengguan*'s patrol) in the *Gu Cheng* functioned as a political mechanism to link control and power over individuals through spatial activities in the marketplace and tour attractions. Lu's construction of tourism space (e.g., the performance area and fixed stalls) in Longcun contributed to his power over the majority of the villagers, facilitated through the alliance with the village political officials. Foucault (1979) discusses how architectural design (e.g., the panopticon) could effectively bring together hierarchical spatial ordering and the control of individual bodies, and argues architecture exists to "ensure a certain allocation of people in space, a *canalization* of their circulation" (Foucault 1984, cf. Lawrence and Low 1990: 485). The control of space was realized through enclosure and the organization of individuals with spatial boundaries, implied in the built forms designed to control and limit subjects' behavior. Built forms help to define new spaces, represent changing social relations, and transform and maintain new socioeconomic

Table 3.1. Summary of how the decisions, policies, and activities (spatial practices) of the main actors in Fenghuang's tourism space (i.e., the tourism brokers including the political authorities and private developers, and the local villagers) transform conceived geographies (representations of space) into lived experience (representational spaces).

	The Public/Private Brokers		The Villagers
	The Political Authorities	*The Private Developers*	
Spatial Practice (perceived) & Representations of Space (conceived)	Political spatial ordering through spatial zoning and patrol by the *Chengguan*; assist private developers with their tourism businesses Foster reciprocal relationship between private developers and political authorities, and create elite alliance and business network; formalize the informal (e.g., the villagers' selling petty commodities and services to tourists) to facilitate control	Acquire long-term land use rights through informal negotiations at low rates; obtain support from the local political authorities	The practice of "hanging out"; verbal protest; daily quarrels; trespass; block the entrance road for tour buses; clandestinely sabotage the tourism facilities
Representational Spaces (lived)	Vending stalls; a cluster of structures which form a performance center; commoditized land—the use value (e.g., subsistence) competed with the exchange value of the same land for tourism development		

practices (Lawrence and Low 1990). Resulting from the combined political and economic power, the site plans for tourism space in both the *Gu Cheng* and Longcun consisted of clusters of built forms in a particular arrangement, and acted as a structural organization of space serving disciplinary ends. The construction and reconstruction of tourism space served as a symbolic system that both signified and magnified the economic and political domination.

Power produces resistance, and resistance operates within the space of power (Yang 2004: 725–726). As Castells (1983) argues, spatial forms not only express and perform the interests of the dominant class, but also are marked by resistance from the exploited classes and oppressed subjects at the same time. The subordinate classes in Fenghuang may have to embrace the hegemony of tourism planning that created a space of exclusion, but they were also tactically engaged in modulating its outcomes, be it through open confrontation (e.g., daily quarrels) or subtle maneuver (e.g., the practice of "hanging out"). In response to the political regulation on tourism space in the *Gu Cheng*, Mu protested to the county government. And those who were involved in ethnic costume rental and the photography business developed long-term business relationships with nearby photo printing shops, not only for printing the digital photos purchased by their tourist costumers, but also for storing their rental costumes and other necessities. The reduction of the dependence on a fixed location helped them better adapt to the tightened spatial control by the local political authorities. In Longcun, to reclaim the village communal space privatized by Lu for tourism profit, the villagers' daily resistance was demonstrated through their trespassing upon the spatial boundaries marked out by Lu's physical constructions. It was one of the villagers' ways of negotiating the hegemony that permeated their everyday life within Longcun, now a busy tourist village, even though their scattered struggles rarely altered the "socio-spatial hierarchy" (Chatterton and Hollands 2003: 184, cf. Su and Teo 2008: 161) established by the dominant class.

SPATIAL CONSEQUENCES

Spatial practices are hardly neutral, fixed, or static. As Moore (1996: x) argues, "The organization of space is not simply a backdrop to social activity, but is the active and interactive context within which social relations and social structures are produced and transformed." Spatial practices act as a medium through which social power is produced and reproduced; power relations are always implicated in spatial practices, and "spatial practices derive their efficacy in social life only through the structure of social relations within which they come into play" (Harvey 1990: 223). Placing spatial practices center stage in touristic power relations, Fenghuang's case demonstrates how spatial ordering enacted practices, through which the tourism brokers as the elites overpowered the villagers, appropriated space, and managed it to organize profit-seeking activities. This in return reinforced their power, enabling new kinds of social domination and exploitation.

Fenghuang's case also reveals how spatial and architectural reconfigurations reflect, and more importantly, transform the local politics. The local political officials' personal wealth and influence were strengthened through their facilitation of private developers' acquisition and transformation of profitable space. As Lew (2007: 152–153) points out, "the division between government, politics and business is more blurred in China than in many Western market economies creating a particular type of power relation often based on elite alliance or business 'network.'" Such elite alliance and business networks at various local levels are a major force shaping physical and social landscapes in late socialist China.

Fenghuang's case further illustrates Zukin's (1993: 267–268) discussion on space and markets, by contextualizing it in a tourist destination. "Command over spaces and times is a crucial element in any search for profit" (Harvey 1990: 226), which is evident in Fenghuang's case. Rigid tour itineraries are fabricated as a "time-space ordering device" (Giddens 1990: 20) to structure the perceptions, interactions, and belonging or alienation of both the toured and tourists; and the temporal-spatial mediation of tourists' consumption affects the benefit distribution among social classes. In Fenghuang, lost in the elites' pursuit of economic profit were the disempowered local lower-class Miao, who were, despite their resistance, subjected to macrolevel planning and decision-making that transformed the fabric of social life in local communities.

NOTES

1. In her study on space and gender among the Marakwet of Kenya, Moore (1996) argues that the organization of space, instead of simply being a backdrop, provides active and interactive context to create and transform social relations and structures. To examine the modern experience of a certain kind of space in China, Rofel (1997) discusses how factory architectural design induces effects of power to shape subjectivities in the industrial discipline of urban factory work; Zhang (2001a, 2001b) explores the reconfigurations of space and power among the "floating population" as a result of massive internal labor migration; and Yang (2004) elaborates modern spatialities of power through the struggle over religious space in a southeast rural community.

2. While drawing inspiration from Lefebvre's (1991, 2004) work, Gatrell and Collins-Kreiner (2006) investigate the case of Haifa's Bahai Gardens in Israel to understand the

distinct spatial practices that enable the tourist and pilgrim to negotiate a shared space; Teo and Leong (2006) analyze the sociospatial politics that make spaces the sites of inclusion/exclusion for the Caucasian and Asian Backpackers in Khao San of Thailand; and Edensor and Holloway (2008) explore the rhythmic qualities of a guided coach tour of the Ring of Kerry in the west of Ireland to understand tourist space, practices, and landscapes.

3. I have discussed elsewhere (Feng 2008b: 211–212) an earlier case of such: the conflict between the YDCC/PATT and the local boatmen over the use of the part of Tuo River running through the central historic downtown for the boat sightseeing business. The conflict had remained unresolved for years despite the county government's mediation. Currently the local boatmen were officially excluded from the central river space where the scenery was the most desired by the tourists.

4. Some of the material in this section appeared in "Space, Power, and Tourism: Notes from the Field in Rural Ethnic China," *Anthropology News* 53(9): 6–7.

Chapter Four

From Tourism Marketplace to Village Homes

Gendered Work among the Local Miao

One day in the *Gu Cheng*, I ran into the wife of Hai.[1] She came to town the day before to visit their daughters, and was on her way to catch a bus back to Wucun. Their daughters Xia and Yun both moved here from Wucun. I stayed with Hai's family during my first visit to Wucun in 2006, and shared a bed with Xia. At that time, Xia just started her tourist photograph business in the village. I had looked for Xia, since I heard that she got married and moved to the *Gu Cheng*. "You missed Xia last time when you visited us. You want see her now?" Hai's wife asked me, and said, "Xia has her own shop now, not far away from here." She then turned around, taking me to see Xia. On our way there, she told me that their oldest daughter, whom I had never met, was divorced, and worked in a flower shop in Shanghai. They visited her for a month. They tried to sell Miao-style jewelry they brought with them in the streets in Shanghai. "It was easy to sell, but the *Chengguan* constantly came to move us on," she said. "I don't want my son to get involved in this kind of business his sisters and I are doing, too hard work and not worth it!"

During my days in Fenghuang, I often heard the local Miao like Hai's wife talk about their expectations of what kinds of work were suitable for men/husbands/sons, and what for women/wives/daughters. Similar to the perception of the Han Chinese, "capable" men were good at *zhao qian* (seeking money), by either earning stable cash through wage jobs in cities, or making big money through businesses of decent scale. Tourism-related petty commodities and services were considered as too trivial and too feminine, and therefore, not ideal for men to engage. Additionally, similar to many other ethnic cultural tourism destinations around the world, Miao women as the bearers of their traditions were featured as one signature subject on Fenghuang's tourism name card. The images of local women, especially those dressed in Miao clothes, were everywhere in local tourism marketplaces. This chapter focuses on the changes in gender division of labor among the local Miao, especially those brought about by tourism growth.

RATHER THAN A "WOMEN ONLY" APPROACH

Miao, along with other ethnic minority peasants in China, are placed in the lower class rungs of the system and bear the burden of absolute, rather than relative, class exploitation, similar to what Smith said about the Maya in the class system of Guatemala (1995:734). Scholars had applied concepts such as "internal colonialism" (Oakes 1995b) and "internal orientalism" (Schein 1997) to analyze the relationship between the Miao periphery as an ethnic tourism destination and the central Chinese state. Being female and poor and of minority status, the lower-class Miao women therefore fall into the category of the "compounded powerlessness" (Ortner 1995:184).

Gender intertwines with class and ethnicity in unequal power relations that operate in the tourism market (Schein 1997; Stonich et al. 1995; Swain 1993). To better understand the construction of structural inequalities emerging out of tourism processes, it is important to explore the gendered complexities of tourism and the power relations they involve (Kinnaird and Hall 1996). Tourism has played an increasingly important role in the quickly evolving social dynamic in many less developed areas, in which process women's roles however had been relatively under-researched. Not until the early 1990s, has the relationship between gender and tourism become a topic of growing concern in anthropology and other social sciences (Ferguson 2011; Kinnaird and Hall 1994; Sinclair 1997; Swain 1995). There is a noticeable tendency in much of the scholarship to devote its discussion exclusively to women's roles and experience in the evaluation of tourism impacts on their status.

To give "visibility" to women's roles in development, it is appropriate for gender studies in tourism to center their analyses on women. Such analyses (e.g., Gentry 2007; Tucker 2007; Wilkinson and Pratiwi 1995; Yea and Noweg 2000) indicate that alterations in economic structures through tourism have differential impacts on gender relations and household organization in agricultural communities, varying from reinforcing the traditional gender division of labor to transforming it by developing new income opportunities. Relevant scholarship reveals the complexity of gender dynamics in tourism development in that tourism may involve women and men in the destination differently (Swain 1989; Tucker 2007; Wilkinson and

Pratiwi 1995), affect women in different destinations unequally (Morais et al. 2005; Swain 1993), and engage individual women in the same destination variously (Cone 1995).

Nevertheless, as Elson (1995:1) points out, women's issues cannot be tackled in isolation from women's relation to men, and "it is necessary to move on from 'women in development' to approaches that emphasize gender relations." While acknowledging the "remedial" function of a "women only" approach as argued by Swain (1995:247), I emphasize the inclusion of men and their experiences in gender analysis. I argue that in order to fully understand the changes of women's roles and status in the context of tourism, such exploration needs to be situated in the interactive relations between women and men, rather than only focusing specifically on women. What are the changes in gender division of labor resulting from a shift from agriculture to wage labor, followed by the growth of tourism? How do processes of change brought by tourism affect household income generation and influence culturally defined "men's work" and "women's work"? How should these changes be interpreted regarding women's status, especially in relation to men's, under different socioeconomic circumstances?

GENDER IDEOLOGY AMONG THE CHINESE MIAO

Tourism impacts on local Miao women and men are mediated by Confucian morality as well as their own ethnic gender ideology, as the Miao have had aspects of their culture influenced by contact with the Han throughout Chinese history. Traditional Chinese (Han) gender ideology is based on the absolute dominance by the male as opposed to the subordination of the female, captured in the Confucian principles such as "*nan zhu wai, nu zhu nei*" (men are in charge of outside the household, and women are in charge of inside the household), "*nan zun nu bei*" (men are superior and women are inferior), "*xian qi liang mu*" (virtuous wife and good mother), and "*san cong si dei*" (three obediences and four virtues) for women. It is also reflected in old Chinese sayings such as "*jia ji sui ji; jia gou sui gou*" (follow the chicken if a woman marries a chicken; follow the dog if she marries a dog). The patriarchal structure is the central feature of gender norms in imperial China with lasting impact today (see Leung 2003 for a detailed discussion).

Miao gender ideology varies, depending on how much an individual or group has been subjected to sinification. The Han recognized two categories of Miao: *sheng* (raw) Miao, who lived up in the mountains and were barely assimilated, and *shu* (cooked) Miao, who settled in the lowlands living among the Han and were mostly assimilated (Diamond 1995). The

Miao in Fenghuang were historically known as raw Miao, and therefore maintained a more pronounced expression of their own culture. Among the Fenghuang Miao, older persons were generally more resistant to the Han, and they used to discipline their children by saying, "Don't be naughty; otherwise, [we will] sell you to the Han"; younger persons were more receptive and increasingly assimilated by the Han due to their frequent contact with the outside world. With the recent promotion of countryside tours in Fenghuang, the influence of the Han has reached farther into remote Miao areas. Some of the old Miao who did not even speak the Fenghuang dialect (spoken by the local Han) learned some spoken Mandarin through their interactions with tourists. Hai once told me that there was an old Miao woman who came to Tuo River Town to buy groceries. As the seller did not understand Miao and she did not understand Fenghuang dialect, she started to speak Mandarin. People around were surprised, laughing, "You an old Miao nanny speaks Mandarin!"

Despite the influence from the Han, Miao gender norms maintained their own characteristics, and appeared less patriarchical with women enjoying more independence, mobility, and sexual freedom. The Miao strictly practice clan exogamy. Among the Fenghuang Miao, marriage is socially disapproved between those with the same family name, even when the couple are not from the same area (Fenghuang County Ethnic Groups Gazetteer Writing Committee 1997; Fenghuang Ting Local History 2003; Shi 2002). Subject to parental interventions and exogamy restriction, however, Miao, especially Miao women, enjoyed more freedom than their Han counterparts. They were exempt from proper Confucian sexual morals and were perceived as being wild and dangerous (Diamond 1988). They were strong with unbound feet laboring in the fields, and could choose their partners based on romantic love and courtship (Fenghuang County Ethnic Groups Gazetteer Writing Committee 1997; Fenghuang Ting Local History 2003).

Schein (1997:83–84) points out that the Miao young people "shared a finely honed understanding of a system that gauged different villages according to their desirability as places to live," which especially concerns the women, as they "have to uproot and accustom themselves to living in a new place." Poverty almost always corresponds with remoteness. The features that matter to a village's popularity include its size, distance to marketplace and urban center, proximity to motor road and bus routes, and availability of electricity and running water. In Fenghuang, a Miao man in a remote village could experience hardship in getting a wife, especially as the cash value of bride wealth as well as the wedding expenses have increased dramatically during the past few decades. In 1990s, a few thousand yuan might be enough for him to get married; nowadays, the estimated cost is likely well over tens of thousands yuan.

Patrilineal and patrilocal, the residents in the same Miao community (except the in-marrying women) in Fenghuang are related through patrilines who normally share the same family name. It is the youngest adult son's responsibility to live with and care for the parents, a custom distinctly different from the primogeniture commonly practiced among the Han. This practice was also described by Shi (2002: 149) and Rack (2005: 35). When the youngest adult son gets married, he and his wife remain living with his parents, and he receives a larger share of the patrimony as compensation for his service, plus the silver jewelry collection from his mother. Miao women receive silver jewelry from their husband who inherited it from their mothers as an important traditional form of bride wealth. The married Miao women could add more pieces into the collection throughout life, and then pass them on to their sons to be used as bride wealth again. Older adult sons set up their own residences with the help of the family to build new houses in the same village, sometimes in lots adjacent to their parents. The relationships and cooperation between households are based on the kinship networks among the male household heads, as the wives are not native to the village. The cultural expectation is that men are the leaders of and the primary providers for the households, and, as Lee and Tapp (2010:161) point out, "men's social roles are often seen as more important than those of women."

GENDER, TOURISM, AND THE MIAO IN FENGHUANG

Miao gender ideology affects, and at the same time is affected by, the Miao women's and men's involvement in the local tourism industry. Galani-Moutafi (1994:126) states that, "gender—namely, the cultural representation and the practice of being 'female' and 'male' in the family and the work place—is a very important dimension for studying social and economic change." Gender, intertwined with class and ethnicity, is lived experience (Moore 2006). There is no better way to grasp the complexity of the dynamic of gendered tourism than to follow the lives of those men and women. The following presents a collection of personal realities of those involved in tourism and, therefore, whose lives had been influenced by it. I compare the changes of women's work and roles in relation to those of men's for a more complete picture on the gendered dimension of the local informal tourism economy. Nash (2001:xi) mentioned that "the most radical changes are those that have occurred in the settings that were in the past the most marginal to the centers of power." By focusing on the local lower-class Miao, Fenghuang's case depicts the relationships of gender to those changes in China's non-Han periphery, and offers a microcosm of much of life in contemporary rural ethnic China and other rural communities worldwide.

"Women's Work" versus "Men's Work"

With the integration of peasants throughout the world into market economies, many rural communities have undergone significant change (Nash 1994). For many of them, the quality of life under new economic circumstances grew dependent on their ability to make money, for which they had to try a mix of strategies (Wolf 2001). In Fenghuang, tourism presumably provides the Miao peasants opportunities for non-agricultural income-generating activities, especially for their small-scale entrepreneurial endeavors, as they require much less capital outlay, no special skills, education, or government permits. However, many of those Miao peasants with ambition to operate their own informal businesses were forced to drop out of the tourism-with-a-decent-profit picture with the soaring prices driven by the local tourism market.

Many in Fenghuang complained about the rapidly increasing prices and rising living expenses. On her way taking me to Jishou to catch an early morning train, a taxi driver chatted about high prices in Fenghuang. She said, "It is much higher than Jishou: red chili peppers are sold at one and half yuan per *jin* in Jishou, but three yuan in Fenghuang." The *liang fen* (cold jelly) vendor in the *Gu Cheng*, whom I frequently went to for my favorite summer treat, had to raise the price from one to two yuan, saying the sugar price jumped from two to five yuan (per *jin*). And there was a small noodle shop across the street from the county government, where I went once in a while for a quick lunch: seven yuan for a bowl of rice noodles with meat, and four yuan for those without. The first time I was there, I was surprised by how little meat came with my noodles. "*Laoban* (means "boss," a polite way to address a business owner), could you please add two yuan worth of meat?" I asked, hungry from conducting interviews all morning long. "Two yuan won't do; three yuan at least," he replied. He soon returned my noodles, with only a couple more tiny pieces of meat. "Why not a little bit more?" I complained. "The prices in Fenghuang have increased a lot, more expensive than Shanghai. The meat now is 15 yuan per *jin*, and even the bok choy is four yuan per *jin*. How much more meat you expect for just three yuan?!" He continued, "The rice noodle shops over the Rainbow Bridge ask for 10 or 14 yuan for a bowl, as their monthly rent is at least more than 3,000 yuan. Mine are cheap. They do business with tourists; I do business with our locals."

With the increasing prices, fewer and fewer of the local Miao could afford the rocketing rent for a shop or stand in a decent location to conduct business with tourists. They had few options other than being engaged as mobile street vendors of petty commodities and services, and Miao women were more likely than men to embrace the new economic possibilities in tourism. Mobile street vendors were predominantly women, selling small souvenirs, flower wreaths, and photographing tourists in rented ethnic costumes. In Xiangxi, customarily, women carried baskets on their back, and men used shoulder poles or pushcarts (Shi 2002). While the reasons were unclear other than tradition, this practice was still being observed in Fenghuang today. Men were rarely seen as peddlers, carrying baskets on their back or in their hands, to walk around and solicit tourists in the street. Some common involvement in self-employment for men includes working as rickshaw pullers, sightseeing boatmen, drivers, and running family businesses (e.g., restaurants, and guesthouses). Other tourism-related activities, such as working as show performers employed by tour developers, were not gender-specific and were open to both young Miao women and men. In Table 4.1, I summarize the typical tourism-related work involving self-employed lower-class Miao in Fenghuang.

"Work is gendered," as Elson states (1995:1), and "some tasks are seen as 'women's work,' to do which is demeaning for men; while other tasks are 'men's work,' to do which unsexes women." Deemed as "women's work," local Miao women in Fenghuang tend to undertake those tourism-related tasks that are either poorly paid or involve a high level of income insecurity. As it is almost impossible for them to afford small shops or even stalls in the local tourist marketplace, most of them are peddlers carrying baskets and wandering downtown where tourists congregate. These tasks are disdained by the men as "women's work," and the women themselves disdain the men who are engaged in the "women's work," based on the assumption that they are incompetent and incapable of securing "men's work."

A common perception among the local Miao (Han as well) is that a family's economic well-being depends on the efforts of its male head and a family's social reputation is evaluated accordingly. A woman street sweeper in the *Gu Cheng* said, "Men should make *da qian* (big money), and income from these kinds of small business is unstable, only suitable for women." Liang expressed a similar opinion when she commented on the fact that those doing ethnic costume rental and photography were almost all women (Photo 4.1). "One or two men might occasionally show up doing this on weekends when there are more tourists," she explained. "Men need work for more secured income." These comments reveal defined notions of independence and masculine integrity from local women's point of view. Men should be the breadwinners in the family, behind which conviction lays the cultural perceptions giving primacy to male economic performance.

Despite being a less stable source of income, the "women's work" otherwise possesses flexibility that allows women to incorporat it in their routine domestic chores. For the camerawomen like Liang, they dropped off their young children at kindergartens or turned them over to the grandparents before leaving for the tourism sites, and picked them up on their way home after that day of business. When the kindergarten was closed on weekends or no

Table 4.1. Gender segregation within the informal sectors* of the local tourism market that typically involve self-employed lower-class Miao.

Categories	Women	Men	Stability, Flexibility, and Visibility	Gender Ideology
Souvenir Vendor	Middle and old-aged women	Very few	Generally unstable and insecure; less profit margin; more flexible; increasing visibility in public space	"Women's Work"
Flower-Wreath Hawker	Women at all ages	School-aged boys, very few adult men		
Costume Rental and Photography	Young and mid-aged women	Very few		
Tour Guide	Young and mid-aged women	Very few		
Rickshaw Puller	Very few	Adult men	Relatively more stable and secure; more profit margin; less flexible; less increased visibility in public space	"Men's Work"
Sightseeing Boatman	Very few	Adult men		
Driver	Very few	Adult men		
Family Business Owner	A few	Adult men		

* The informal sectors considered here reflect those self-employed tourism activities. Note that some of the categories (e.g., tour guide, sightseeing boatman) include both self-employed opportunity and wage employment. Also note that these informal sectors do not only involve the Miao, but also the Han.

Photo 4.1. A cameraman was taking a break in the Culture Square in downtown Tuo River Town. He was the only cameraman among the many camerawomen in the square who were busy dressing up their tourist customers and taking pictures of them in front of the phoenix sculpture. (Photo by Xianghong Feng, 2014).

extra help was available, they either stayed at home or took the children with them to the marketplace. With the responsibility of finding a balance between earning extra income and fulfilling their traditional roles, such work, though flexible, burdens the women whose regular workload includes housework, childcare, and, sometimes, agricultural production in the field, especially when the husband works away as a migrant laborer.

A Miao woman, in her early thirties, sold flower wreaths in the *Gu Cheng* during the summer of 2010. She arrived in early mornings when tourists were still asleep. Temporarily settling down on a street corner, she started making flower wreaths with freshly picked flowers from her village. The finished ones were carefully placed in a small basket. She then carried the basket around to sell them to tourists for a few yuan each. Her husband was a migrant laborer. She had to take care of the crops, with the help from her parents-in-law. Summer was an ideal time for her to sell flower wreaths, as she relied on her ten-year-old daughter who was on summer break for the household chores, especially taking care of her three-year-old son.

From Peddler to Shop Owner: A Miao Woman and Her Family

Xia was Hai's youngest daughter. Hai had three daughters and one son. I first met Hai and his family in 2006 when they were actively involved in tourism in Wucun. He directed the village performance team, his wife managed a souvenir stall, and Xia took photographs of tourists dressed up in Miao costumes rented from her. At that time, Xia had just returned from the city where she worked as a migrant wage laborer. She had a boyfriend named Shun. They met each other back when her family was selling souvenirs in Huangcun village, where he worked as a guard for the PATT. Later, she married Shun, and moved into his family's home in the *Gu Cheng*. She continued her costume rental and photography business for a while, and then managed to rent a small shop front in one of the busy riverside streets to sell clothes.

A small room adjacent to the back of the front shop housed the workshop for a photo printing service. It was equipped with a desktop computer, a printer, and a photo laminator, and furnished with a desk and a couple of chairs. Xia sold clothes in the front room, and Shun was in charge of the photo printing business in the back room. Most of the time, she was busy, while he enjoyed playing video games and watching online TV episodes on the computer, except when interrupted by the camerawomen who took tourists in. Shun displayed all the digital photos on his computer screen for the tourists' selection of those to be purchased. The photos selected were to be printed and laminated right there. Normally the tourist paid ten yuan per photo, with nine yuan going to the camerawoman and one to the printing agent. Unlike most of the printing agents who always tried to promote sales, Shun went through the photos quickly—neither did he pay praise to any photo, nor did he leave enough time for the camerawomen to convince their customers to purchase more. While displaying each photo, he asked, "Want it or not?" He then clicked the mouse, quickly moved on to the next one, sometimes even before the tourist could make a decision, and asked again, "Want it or not?" To Shun, the potential profit from this petty business was not significant enough to turn him away for long from his electronic diversions.

The longest business hours for Xia's shop were for summer days. The shop remained open from early in the morning till almost midnight, as the riverside streets were packed with tourists who enjoyed the night scenery and cool air. With little assistance from Shun, Xia managed the entire shop on her own, greeting customers, bargaining, finalizing sales, and going to local wholesale stores to replenish the supplies. Shun and his parents' house, which Xia had married into, was a five-minute walk up the hill behind the shop. Her father-in-law was paraplegic, and her mother-in-law did most of the household chores including cooking. Xia took short meal breaks when the business slowed down, walked back home, had a quick bite to eat, and then loaded up a big bowl of rice and food to bring back to the shop for Shun.

While Hai was proud of his daughter's success from being a peddler to a shop owner, he was not satisfied with his son-in-law. "What do you think of Shun?" he asked me all of sudden during one of our conversations in Wucun. He just returned from visiting Xia and Shun. Before I figured out what he meant, he started to complain that Shun stayed around in the shop doing nothing to assist Xia after quitting his job with the PATT. He suspected that Shun was actually fired rather than having quit on his own. "As a man, how could he be like that?" Hai continued. "Even though he is not willing to help much, at least he shouldn't have turned those customers away with his cold attitude." He mentioned that some camerawomen were upset with Shun's manners, which cost them good sales of photos. "Maybe that is just his personality. At least he treated Xia fairly well," he comforted himself.

One morning I was visiting Xia's shop. At that time she had closed the photo printing business, and switched from selling clothes to handwoven scarfs and vinyl CDs that were popular among tourists (Photo 4.2). Shun stopped by around noon and brought two containers of watermelon and several plastic bags

Photo 4.2. Xia (right) was helping two tourist girls trying on scarves in her shop. (Photo by Xianghong Feng, 2014).

of CDs. We had some watermelon, and Xia wanted to take me to a restaurant for lunch. I was reluctant to cost her time away from the shop. She assured me, saying, "No worries. Shun is here." I remembered Hai's complaint of Shun's cold manner with customers. "That is not true. Shun is actually good at doing business. He had to move on quickly when printing photos, as there were other customers waiting. How could he spend too much time with one customer?" Xia said, in Shun's defense.

Xia was content with her married life. Indeed, for Xia, this was marrying-up. Her marriage had moved her from the countryside to the capital town where the local tourism industry prospered with the large influx of tourists. After all, marrying into a family with a residence in the *Gu Cheng* provided crucial resources for her to succeed in her business endeavor, providing her hard work and ambition. Her plan was to rebuild the current house, and turn it into a guesthouse in a year. Hai and his wife invested their savings in running a guesthouse in one of the riverside streets in the *Gu Cheng*. It was under renovation in the summer of 2011 when Xia was in the process of getting official permits for rebuilding hers. Like his wife, Hai was not willing to let their son engage in petty businesses. He intended to turn the guesthouse over to the son once the business was on good footing, but expected him to maintain his full-time job at the local transportation bureau. It was a decent job with, however, a modest salary. "My daughter-in-law is also running a small business. So I told my son that he should work harder, and be a capable man, otherwise, his wife might not want to be with him if she made more money," Hai said.

Woman-Woman Tie in Tourism Marketplace

Marketplaces could serve to unify and make visible women who might otherwise be dispersed and hidden in their individual households, as in the case of the Maya women vendors in Guatemala (Little 2004:169). Local women in Fenghuang continued to express culturally specific expectations about men as the chief economic providers for the family; nevertheless, they came to depend more on other women, kin or non-kin alike, for cooperation and assistance. In addition to the mutual benefits directly concerning their businesses, these Miao women's social networks formed in the marketplace provided opportunities for the exchange of favors and information in daily life beyond the marketplace.

1) The Two Sisters Xia and Yun and Their Friends

When Xia started her own shop, she turned her photograph business over to her *er jie* (second older sister) Yun, whose husband was a migrant laborer.

Yun became friends with several other camerawomen, and they gathered around the tour attractions to seek potential customers. One of their favorite locations was in front of Xia's shop, as it was in one of the riverside streets, with popular scenic spots for photography and a steady influx of tourists. They dropped their baskets which stored the rental costumes at the shop, hung a digital camera around their necks, and held plastic laminated brochures in hand on which were displayed sample photographs. They approached tourists and actively sought those interested. They dressed them up in ethnic clothes and headdresses, chose background scenes, directed their poses, and took as many photos as possible. They took their customers to the back of Xia's shop to check out the photos and to print those selected for purchase. When having a break, they sat at the steps of the shop, made sure not to block the way of the shop customers, switched from speaking Mandarin to Miao language, exchanged their daily business stories, chatted about their family and children, and played jokes on each other. They still paid attention to the passing tourists, so as to not miss any potential customers (Photo 4.3).

Xia, Yun, and their camerawomen friends formed a cooperation of business and coalition of support for each other in the local marketplace. For Yun and her camerawomen friends, it was important to develop a long-term business partnership with a nearby photo printing shop such as Xia's. In doing so, they could store their baskets with rental costumes in the shop, which increased their mobility to approach more tourists; they could, immediately after shooting the photos, take their customers there to choose and print photographs without the hassle of looking for a shop and negotiating the printing price every time. Xia benefitted because her photo printing could receive a relatively stable business from Yun and other camerawomen.

When the renovation on her parents' guesthouse was completed, Yun quit being a camerawoman and helped around at the guesthouse. There were a handful of guesthouses along the same street. Yun and the girls from the next-door guesthouses became good friends. In the morning, they went to the temporary long-distance shuttle station on the other side of the street, to solicit tourists who just arrived in town. If one of them had enough guests for that day, she would solicit tourists for the others. During lunch or diner, they often loaded their bowls with rice and dishes and visited with each other. Whoever cooked a nice dish, the others would take an empty bowl to go eat. One day Xia and I went together to visit her parents at their guesthouse. Yun and her friends were just back from the station. "We saw foreign tourists today," Yun said to us. "Too bad that we can't speak any English!" I offered to write down a few expressions in English for them to show the foreigners. On the notebook Yun handed me, I wrote: "Hello." "Are you looking for a place to stay?" "Our guesthouse is in the *Gu Cheng*, and I can show you the room at no charge. If you decide to stay, 120 yuan per night." "Have a nice

Photo 4.3. Yun (foreground in jeans) and one of her fellow camerawomen were folding their Miao costumes and putting them away in front of Xia's shop, after a group of tourists rented them to wear for photographs. (Photo by Xianghong Feng, 2011).

trip." They tried to learn how to speak them. At the end, they used their cell phone to record me. "Surely by tomorrow we don't remember how to say it anymore," they giggled and said.

2) Liang and Her Camerawomen Friends

Also a camerawoman, Liang did her business in the Culture Square. She was a Han from rural Huaihua. She was in her mid-twenties, and almost illiterate. She had worked as a migrant laborer, and her first husband died from an accident in the same factory. She moved to Fenghuang after marrying Hong in 2009. Hong was eleven years older than her. He was from Liaojiaqiao Town, and self-employed as a private contractor building houses in Tuo River Town. They rented a room close to the *Gu Cheng*. While Hong spent long days with his construction team at work, she either stayed at home taking care of their one-year-old son, or came to the Culture Square to make extra money when their son was at the kindergarten. When I first met Liang in 2011, she bought a used Nikon digital camera online at the price of about 3,000 yuan, and started to practice photography. "Quite easy," she said, "and the composition is the key. I practiced myself, and other people taught me some tricks too."

Her income was unstable: sometimes she made a couple hundred yuan within half a day, but some days she ended with no business at all. She wanted to make enough money to cover her son's kindergarten cost. It was 150 yuan per month along with a registration fee of 400 yuan in 2011, and in 2013 it increased to 220 yuan per month and 1,000 yuan for registration. "Despite the cost, my husband and I thought it is a good thing to send our son to kindergarten as he will learn more there," Liang said to me. Whatever she made was a nice extra, as Hong's earning was what the family depended on. Once at the dinner table, Liang joked to me about how she asked Hong for money to buy clothes, and he was not willing to give her any, saying that there was not much point for getting new clothes as she would just make them dirty while working.

As a non-native of Fenghuang, Liang seemed gradually adapted to the new place over the years. She got along well with her sisters-in-law. On her business days, Liang took a lunch box, or counted on her sister-in-law to bring her lunch. She made friends with other camerawomen, and the Culture Square was their common business territory. They might happen to be soliciting the same tourist at the same time. Then it was up to the tourist; there were rarely disputes. If the tourist picked Liang, but liked another camerawoman's rental costumes, the other woman would loan Liang the costumes. Liang could share with her the money she made from this tourist. But most likely it was not necessary, as they often borrowed one another's costumes.

They hang out in their spare time as well. They once went to the Nanhua Mountain in town, which the county government newly leased to a developer to run as a park. The local residents could visit the park for free. Liang told me, "I couldn't speak the Fenghuang dialect well, and knew little of Miao language. So the park wouldn't let me in at first, until my friends assured the park staff at the ticket checkpoint that I was indeed local as I was married to a local." As a Han woman who was a recent immigrant to Fenghuang through marriage, the ties formed in the marketplace with other local Miao women provided Liang her own supporting networks and access to local resources and opportunities that would be otherwise be difficult to obtain. When Zhao joined them, Liang and other camerawomen offered her the same comradeship. Zhao's son was about the same age as Liang's. Their sons attended the same kindergarten, as Liang introduced it to her. Liang always recommended the restaurant where Zhao worked to other people who wanted to eat out. I once took Liang and Hong out for lunch at Zhao's restaurant.

3) Hua and Her Tour Guide Friends

Hua was my Miao language interpreter for two summers. She is Miao from Lacun village in Sangongqiao Township. When she was ten, her father died unexpectedly from an acute illness. Her mother decided not to remarry, and left for Hangzhou to seek wage jobs to support the family. Hua and her younger brother were raised by and lived with their paternal grandmother (a widow since her middle-age) in the village. I met Hua at Jun's travel agency as mentioned in the Introduction and Chapter 1. Hua was in her late teens, and then worked as an intern tour guide there, upon finishing her education at a local vocational school.

Hua made friends with many other tour guides (typically young Miao girls) working for either travel agencies or tour developers. When they took tour groups to the local attractions, they often ran into one another at the same site. They asked about each other's tour group for that day during quick passing-bys; they told the stories of their best and worst tourist customers while waiting for their tour groups to finish watching shows or taking pictures. These casual interactions gradually developed into more formal networks, upon which they started to socialize outside the tour sites, including dinner at restaurants, birthday-celebration parties at the karaoke bar, and shopping on market days. They shared the information about their earnings, their tricks dealing with difficult tourists, and their experiences with the tour agencies or developers they worked for. They sometimes introduced potential guests to each other. They carefully cultivate these ties as a readily available resource for reciprocal favors.

Hua stayed at Jun's agency after her internship. But she was not ready to settle down yet, despite her appreciation for the apprentice opportunity and guidance offered by Jun, a friend of her uncle. Hua gathered information from her tour guide friends, and actively sought better offers. She once seriously considered another travel agency, as the base salary was considerably higher. Before she made up her mind, she spent a few days joining the tour groups led by her tour guide friend who worked for that agency. Hua decided to remain at Jun's agency after that, saying, "It was too rigid schedule and much harder work over there!" Without the opportunity provided by her friend for a firsthand experience, Hua might not be able to tell if it was right for her. She told me, "If I went to the other agency to find out only later I didn't like it, I was not going to come back to Jun's. Otherwise, I would lose my face."

Men and Women in a Miao Tourism Village

With the transition from agriculture to wage labor and tourism, village household organization has undergone changes, and men and women have been affected differentially by the new economic circumstance. Galani-Moutafi (1993) discovered in a Greek island village that as young men turned away from agriculture to migrant wage labor and men's orientation moved away from agriculture, women became the actual owners of property/head of the household. She (1993:258) further pointed out that during the transition from agriculture to tourism, women became the backbone of most family enterprises in tourism. In rural Fenghuang, however, this may not necessarily be the case.

In rural Fenghuang, for those households that both husbands and wives migrated out as wage-laborers, the paternal grandparents took care of the crop fields and their grandchildren, and the migrant couples' responsibility was to send enough money back regularly. The left-behind old and young could benefit little from the tourism. For those households with only husbands away as migrant workers, the husbands were the primary source of income while the wives, sometimes with the help of their parents-in-law, managed everything back home. For those households in which both husbands and wives stayed, the husbands normally were the planners and decision-makers of tourism-related family enterprises.

In any of the above situations regarding the migrant pattern, men consistently appeared to be the financial backbone of the household, despite women's increasing participation in tourism-related activities. The changing socioeconomic circumstances resulted in changing gender division of labor, and women contributed to household income as well as gained an increased role in household expenditures. However, the economic control seemed to be remained in the hands of the men, as women's contributions to income were considered insecure and insignificant. Women's labor as producers of goods and services within the household was likely taken for granted and counted as free. For those who were engaged as part-time mobile vendors outside their household chores, their cash income was neither as stable nor as much as their husbands' migrant wages or profits from small business; their income also appeared less visible as it tended to be used as extra money to subsidize small household expenditures on a daily basis before accumulating into any larger amount.

1) Jing's Family

Longcun was among those local Miao villages in which everyday tourism-related activities were shaping a new socioeconomic order. In Longcun, even though women increasingly participated in family enterprises in the local tourism market, men made strategic plans and directed decisions at the macro-level while women, if they were involved, remained as the manager and care-taker of daily practice at the micro-level. For the households that ran family restaurants to provide standard meals to tour groups, the men were in charge. Jing was one of them. I once invited San Ge and Jing to Beijing, as they were longing to visit Tiananmen Square, especially the Chairman Mao Memorial Hall. I thought I could show them around for a few days in Beijing, and then leave with them for Fenghuang. San Ge told me that it was very unlikely that Jing could make it, saying, "He is in charge of the restaurant. He won't feel comfortable to leave it to his wife. Not even for a few days."

Jing's wife prepared regular meals for the household and family visitors, independent from the large meals catered to tour groups. Because of the restaurant business, her cooking workload was reduced as there were often leftovers that only required reheating in the wok, or adding a few ingredients. Even though Jing's wife was not involved much in the restaurant business and her household chores around the kitchen seemed to be significantly less, her workload was heavier. As Jing was tied to the restaurant and a schedule determined by the itineraries of various tour groups, she was the one who took care of the farmwork.

Jing's wife was not alone. Near the show performance ground in the village, there were three middle-aged women vendors, among others, selling souvenirs. Their husbands had all migrated to Hangzhou as wage laborers. They set up their stalls every morning before the first tourist group's arrival and stayed until noon or early afternoon when they had to go to work in the field. With their husbands' year-round absence, they had to take care of all the agricultural work and domestic chores, and sometimes their elderly parents-in-law. The return was the monthly remittance from their husbands. One of them said, "My husband sent back 1,000 yuan every month. I wanted him to send back more as everything is getting expensive here nowadays. But I know his life there is tough too as I visited him before. After what he spends on rent and food, he doesn't have much left."

2) San Ge's Family

Even though Jing controlled the income from the family restaurant, his wife had significant influence on his decisions, especially money-related issues. San Ge was Jing's youngest brother. Migrated back from Hangzhou, where he had worked for eight years as a wage laborer for various factories making drive shafts, San Ge joined the minivan rental business for about half a year before selling his van due to the diminishing profit. San Ge's eight-year-old son had an accident and was hospitalized for two weeks including a week in the Intensive Care Unit (ICU). It cost him more than the money he got by selling his van (more details in the next chapter). Jing loaned him 4,000 yuan to pay off the hospital bills. He soon borrowed money from his friend to pay back Jing, as he sensed Jing's wife's reluctance.

Women's labor was indispensable for an ideal household. One day the head of the village production team was hosting two friends from Tuo River Town. He cooked a rooster, and invited Qiu, Mi, San Ge, and me to join them for lunch. I was the only woman among them. Noticing that I did not eat much, they teased me, saying, "Those women who don't eat much can't find a husband here. Prettiness is less important. If she does't eat much, she won't have the strength to do things." Another day when I was chatting with several women, they commented on how good-looking San Ge's wife was, and one of them said, "What was the use of prettiness? Look at his sister-in-law [Jing's wife], not pretty at all, but helped her husband run a good household!" Women's roles were viewed as significantly complementary to the men's. San Ge's wife left him and their young children while both of them were migrant workers in Hangzhou. San Ge said, "My older brother's situation is much better than mine as he has a wife, and that they both work to support a family is way easier than me raising my two children alone. In addition, as the youngest son, I have to care for our elderly sick mother." He continued, "I can't afford to get another wife. It costs tens of thousands yuan." He was not interested in remarrying any time soon, and said, "I just want to save more money. My two kids are growing up. After the elementary school, it will cost much more."

A similar pattern was observed in Qiu's household. He was the first in the village to open a family restaurant, and he started a guesthouse business in the *Gu Cheng*. Qiu's sister-in-law worked in the guesthouse full-time, while his wife remained working around the household, as did Jing's wife. I asked Qiu, "Does tourism have any impact on women's status within the household?" Qiu replied, "No impact. For example, most of the vendors in the village are women, and they can more or less make some cash from the tourists daily." I pursued the question further, "Will that make what they say matter more?" Qiu asserted, "Not really." He continued, "We used to have very few income sources. Before tourism, women in our village collected herbs in the mountains, and sold them to the middlemen on market days. They made five or six yuan a day. Now as vendors, they are happy if they could make ten yuan a day." "Do they need to turn in the money they earn to their husbands?" I asked. Qiu answered, "No. They keep it and use it for things like children's school expense, salt purchase, and electricity usage."

INCREASING FLEXIBILITY OF
WOMEN'S ROLES: AN EMPOWERMENT?

The above discussion portrayed individual Miao women and men in the marketplace and within their households, whose stories shed light on the dynamics of gender roles under the transition from subsistence to cash economy. In the subsistence economy, men were the main labor force in agricultural activities; the domestic chores largely fell onto women's shoulders. In the cash economy, men's main responsibility was to become the stable source of household income through working as wage laborers (e.g., the husbands of Yun and the women vendors in Longcun) or running small businesses (e.g., Jing, Qiu, and Hong); women often sought opportunities for supplementary cash income in addition to their domestic chores, and some of them (e.g., Jing's wife and the women vendors in Longcun) had to undertake agricultural work if their husbands were not available, either away as migrant wage laborers or busy running small businesses.

This indicates the flexibility of the gender division of labor under changing socioeconomic circumstances. It is important to note that the flexibility occurred mainly for women to undertake "men's work" (such as agricultural work), not for men to engage in "women's work" (such as domestic chores and petty commodities and services). This flexibility for women is to maintain their subordination to the men under the changing socioeconomic circumstances, in which men's main responsibility shifts from agricultural production to *zheng qian* (earning cash). The role of sacrifice for the collective good to realize "ideal households" falls inherently on women rather than men. The traditional gender ideology of men's superiority is therefore not challenged but reinforced with the penetration of market economy. With the diversification of the rural economy attracting male peasants away from the land and women replacing them in the fields, some might conclude that women experience more autonomy as presumably reflected in the increased flexibility of their roles. But, as pointed out by Davin (1995:37), women "take over only when farm work has become less desirable to men than some better-remunerated alternative."

Scott (1995:115) states that men are most likely to move upward to improve their positions in the labor market, while women had much less mobility and "such movement as did occur was likely to be a lateral movement between unskilled jobs rather than an upward career progression." The high profile of women and their visibility in the tourism marketplace might not challenge the traditional gender ideology, but paradoxically reflect the lower status of women. In Fenghuang's case, the increased flexibility of women's roles was a result of the reality that Miao women, who occupied the low end of every form of power in the local and national social structure, were being pushed around by other more powerful internal and external agents. Thus, the increased flexibility of women's role is not necessarily an indicator of women's empowerment. Any assertion concluding women's empowerment from the increased flexibility of their roles without a critical analysis of its relations to those of men's is an incomplete view: women could appear more autonomous when compared to themselves at a different time, but might not be so when looking at their relative positions to men.

When comparing women with men, in terms of expanded earning power derived from work linked to tourism, men enjoyed greater access to individual income accumulation than did women. However, this is not to deny that women, as participants in economic and social transformations, actively challenge sociocultural rules on gender roles under these transformations, as demonstrated in the case of Xia and Hua. Xia took advantage of her relocation (through marriage) from her home village to the capital town, and almost single-handedly started and expanded her small business. Hua and other Miao tour guide girls, as the younger generation, were ambitious in pursuing their economic independence.

Furthermore, women's contributions were socially valued for the maintenance of an ideal household, which the Miao women, unified with their men, worked hard to maintain through a flexible "sex/gender allocation of labor" (Du 2000) contingent upon the changing socioeconomic context. Supporting Jing's commitment to the family restaurant business, Jing's wife assumed almost all the other work, and their joint effort maintained a good household as perceived by their fellow villagers. In contrast, the fellow villagers expressed sympathy for San Ge's situation, as his wife's departure left him with a difficult job of taking care of the family.

When compared with themselves in pre-tourism times, those women involved in tourism-related work enjoyed relatively enhanced status, increased flexibility in their roles, decreased economic vulnerability, and increased autonomy. All resulted from their cash income (despite its insecurity and insignificance), as well as a shift toward an emphasis on female-female ties (e.g., among the local camerawomen, the guesthouse girls, and tour guide girls) for kin and non-kin alike in marketplaces and their home villages. Additional benefits include an apparent increase in confidence and an expansion of horizons through daily contacts with tourists and other outsiders involved in the local touristic system.

Through Fenghuang's case, I emphasize two general points regarding gender analysis in tourism: first, in order to fully understand gender dynamics in tourism, it is important to contextualize such analysis with the particular historical and sociocultural factors in a given locality and against the backdrop of the global economic trend; second, for a fair evaluation of women's role and status change, such exploration needs to be placed in the context of interactive relations between women and men, rather than only focusing specifically on women. Swain (1995:264) points out that the study of gender is "both a scholastic and political endeavor," and "one of the avenues to understanding the dynamics and promoting change toward equality is through the study of gender relations." As a common development strategy, tourism presents both opportunities and challenges for gender equality and women's empowerment (Ferguson 2011; UNWTO and UN Women 2011). The two general points highlight a more holistic approach in the study of gender relations in tourism as well as in other development processes. Such an approach provides clearer understanding of women's relative position to men under changing socioeconomic circumstances.

In the context of economic change through tourism, local lower-class Miao families in Fenghuang had to work harder and intensify the use of their remaining resources and assets, particularly the labor of women, to maintain necessary levels of household income. This intensification of work placed an unequal burden on women. Miao women, apart from seeking uncertain income opportunities in the tourism marketplace, saw their domestic chores and agricultural workload become greater than before due to the increasing unavailability of their husbands. As Gonzalez de la Rocha (2001) points out, the accumulated evidence suggests that women have endured a heavy share of the social cost of economic change (Beneria and Feldman 1992), and that patriarchy (rule by men) is a durable feature of capitalist economic development (Cockcroft 1998).

NOTE

1. This chapter is revised from "Women's Work, Men's Work: Gender and Tourism among the Miao in Rural China," which appeared in *Anthropology of Work Review* 34(1): 2–14.

Chapter Five

Prosperity for Whom?

Tourism and the Poverty of Resources

"Can you help me think what small business would be good for me to do?" San Ge asked me.[1] It was after dinner at Jing's house, my host family in Longcun village. It was one summer evening in 2011. He just returned from Tuo River Town, where he drove tourists with his minivan to the countryside with no pubic transportation. Business had been bad again that day for San Ge. "I made only 20 yuan today, not even enough to cover the gas money," he complained. I was not able to answer San Ge's question right then. He sought my advice, and believed in me as an "expert" in researching Fenghuang's tourism who could point him to practical alternatives. It was not easy to offer a satisfactory answer to his question, a question at the core of many issues concerning Fenghuang's tourism development. Why was it so hard for the local Miao peasants like San Ge to find even modestly profitable business opportunities under the "prosperity" of Fenghuang's tourism?

I discuss in previous chapters the exploitative and exclusive nature of Fenghuang's large-scale and elite-directed tourism development. In this chapter, my focus shifts to the exploration of this nature from the experiences of the local Miao peasants to see how they, as those who are on the bottom of the local social hierarchy, are living with it. As an effort to investigate the relationship between agriculture, migration, and small-scale entrepreneurs among the peasants, I ask the following questions: What kept the peasants going as migrant workers? What motivated them to engage in small-scale entrepreneurial endeavors? What enabled them and constrained them? What they were going through sheds light on the social reality faced by many Chinese ethnic poor like those in Fenghuang in the context of the tourism market economy, in which they had to struggle to live off the land and seek security beyond agriculture.

LEAVING AGRICULTURE

Since the establishment of the People's Republic of China in 1949 throughout the Great Leap Forward, Cultural Revolution, and Reform and Open-
Up eras, the Chinese government has maintained the tradition of directing
socioeconomic change through the use of motivational political slogans that
help inform and promote compliance with official policies (Li 2011). In the
early 1980s, China's rural development strategy promoted township-village
enterprises (TVEs), well described by the slogan "leave the field but not the
countryside; enter the factory but not the city" (*li tu bu li xiang; jin chang bu
jin cheng*) (Ho 1995). Since the 1990s, reflected in the new slogan "leave the
field and the countryside" (*li tu you li xiang*), rural-urban migration in China
had been encouraged and facilitated by the local governments, aiming to absorb the surplus rural labor and improve the rural livelihood without prior or
large-scale investment (Liu and Chen 2007; Song 1996; Xie and Xu 1988).

The study conducted by Croll and Ping on eight villages in four different provinces of China suggests that under the range of conditions in which
migration supplements, subsidizes, and substitutes for village agriculture, the
peasants in different villages have concluded themselves "that agriculture is
an unprofitable, unattractive and even redundant economic activity" (1997:
129). Croll and Ping (1997: 141–142) point out the paradox of "leaving agriculture is for agriculture," in which "individual villagers have increasingly
left agriculture by either transferring into local non-agricultural occupations
or migrating out with the express aim of earning cash incomes in order to
subsidize family agriculture production."

"Leaving agriculture is for agriculture" may only reveal a partial reality
in the peasants' preference for non-agriculture activities. Scott (1976: 2)
states, based on the "safety first" principle and the "subsistence ethic," that
the preference of those peasants in precapitalist agrarian societies was not
to take risks for profit-maximization, but to avoid economic disaster, as "a
consequence of living so close to the margin." With the penetration of market economy which is expedited by local policies for rural development, for
many village households like those in Fenghuang, agriculture has become
a part-time sideline that often falls onto the shoulders of the women and
elderly, supplementing the more important non-agriculture economic activities sought by their husbands and adult sons, as discussed in the last chapter.
Under the current socioeconomic circumstances, many of the peasants who
seek security have to look beyond agriculture, to out-migrate as wage laborers
or search for small-scale business opportunities.

Rural migrant workers in cities are widely stigmatized as poor, unintel-
ligent, and uncivilized (Zhang 2001b; Cho 2012), and their jobs as dirty,

dangerous, and demeaning (Tao 2006, cf. Gao and Smyth 2011:164). Knight and Gunatilaka (2008, 2010) find out that the mean happiness levels of rural-urban migrants are lower than both those peasants who stay in the countryside and those urbanites living in cities. Gao and Smyth (2011) explore what keeps the migrant workers going, and suggest that it is an optimistic expectation about their future income that motivates them, despite the reality that their living and working conditions in cities are so harsh. I however consider what keeps them going may be simply the fact that they could hardly afford otherwise: out-migration increasingly became not so much an attractive alternative but a necessary step in order to meet the rising costs of household daily necessities, school fees, and health costs. Their "optimistic expectation" thus would be better described as their expectation of the positive future benefits of their wage savings, including for major household expenditures such as home renovation and children's education, or funds for small businesses.

While wages do make a difference, the greatest disequalizing factor in rural incomes is from rural enterprise or small business (Khan and Riskin 2005). The households that can accumulate capital do better than those who merely diversify into migrant labor (Mohapatra et al. 2006). Zhou et al. (2008) indicate three factors that enhance the ability of households to diversify their income, including the head start, labor power, and involvement in small businesses. As the impact of labor power has diminished according to Zhou et al. (2008), doing small business is the main route leading to higher income for the ordinary peasants who did not have a head start.

FROM "RESOURCES OF POVERTY" TO "POVERTY OF RESOURCES"

One perspective on China's rural developmental process envisions a development ladder in which peasants move up, rung by rung, first as subsistence farmer, upward to migrant laborer, and then to petty capitalism (e.g., Mohapatra et al. 2006). This perspective sees the development of increasing occupational complexity as progressive and empowering. From this perspective, occupations evolve according to the wealth and location characteristic of regions, and "least complex occupations are dominant in the poorest and most remote regions and increasingly complex occupations in wealthier and more developed ones" (Mohapatra et al. 2006: 1035). Wealth and geographic location then are considered with this perspective as the two crucial elements affecting the micro-level evolution of the dominant occupation in rural China.

However, I advocate a different view. I argue that rising wealth and improved local environment (e.g., infrastructure), resulting from increasing levels of development, does not necessarily lead to the proliferation of more complex occupations for the majority poor. As Zhou et al. (2008: 517) point out, the four-stage developmental pattern of occupational succession and differentiation (proceeding from agriculture dominant, to migration dominant, to local enterprise dominant, and finally to industrial employment dominant) proposed by Mohapatra et al. (2006) is a developmental sequence of the modernization type, in which inequalities between communities and among individuals within the same community are "a mere result of the time lag between when they proceed through various steps on the ladder." Echoing Lee and Selden (2006), I consider that they are instead symptoms of a structural system of inequality, in which a few get rich at the expense of the vast majority of China's poor such as those Miao peasants in Fenghuang. With tourism growth, the local economy may seem to provide local residents improved infrastructure and expanded economic opportunities, but it is developed at the cost of the local poor's access to, therefore their equal share of economic profits from, these resources. Thus, their poverty is, I argue, a "poverty of resources."

The concept of "poverty of resources" (the erosion of the "resources-of-poverty" model of survival) is discussed by Gonzalez de la Rocha (2001, 2007) in light of the economic crisis faced by the poor urban households across Latin America. She defines the "resources of poverty" as "the diversity of income sources and to the social organization of households or the social base that makes survival possible" (2001: 76). She refers to the "poverty of resources" as "the outcome of labor exclusion and the persistence and intensification of poverty," and states that it "signals the erosion of the social and economic conditions for survival" (2001: 86). The "resources," which had been crucial to the poor's survival and livelihood under scarcity, include a wide range of tangible and intangible assets, both material (e.g., natural resources such as land and water) and social (e.g., social capital and insurance such as kinship network and reciprocity) (Gonzalez de la Rocha 2007; Scott 1976).

Here I apply Gonzalez de la Rocha's concepts in a rural setting in China, and integrate them with Scott's (1976) idea of "moral economy" in precapitalist agrarian society, to draw a clearer contrast between the "resources of poverty" and "poverty of resources" for the rural poor (Table 5.1). As Gonzalez de la Rocha (2007: 48) argues, "the 'resources of poverty' model of survival has been eroded, and the lives of the poor are better described today by the reverse formulation: the poverty of resources." The following discussion contextualizes the "poverty of resources" in Fenghuang's tourism development, to shed light on the problems that many ethnic poor in China are facing in their struggle to break the cycle of poverty.

Table 5.1.   Resources of Poverty versus Poverty of Resources.

Resources of Poverty	Poverty of Resources
Facilitate survival	Erode the capacities for survival
The right to subsistence	Profit maximization
Enhanced security	Increased vulnerability
Survival of the weakest	Survival of the fittest
Inclusive	Exclusive
More humane	More economic
Moral economy	Market economy

LABOR, CAPITAL, AND TOURISM

Wolf (2001: 257) states, "From the point of view of ensuring their survival, they [peasants] may wish to produce the many different things they need themselves and to reduce their dependence on the market. From the point of view of obtaining money, they will try a mix of strategies that will yield money." The combination of diverse income sources from as many possible members within a rural household serves as a social mechanism to cope with scarcity, and an individual member could switch between different occupations not only at various stages in life, but also engage multiple income-earning activities at the same time (Gonzalez de la Rocha 2001, 2007).

In recent years many rural communities worldwide have undergone significant change as a result of the transition to income-generating activities that are non-agricultural including migrant wage jobs and tourism-related work. In China, increased mobility through migration and tourism has played significant roles in bringing its ethnic peripheral areas into closer and more forceful contact with the world outside them (Chio 2011), and one of the most noticeable changes is the influx of manufactured modern goods. As Moore (1996: 138–140) points out, based on her observation of the Westernization among the Marakwet of Kenya, modern items can be acquired by anyone who has access to the appropriate cash resources, and that the acquisition and display of certain items is, therefore, highly desirable, since this is the only way in which most people can participate in a status system which values the "modern" over the "traditional." The cheap manufactured goods flowing from the outside, at the same time, have outcompeted and nearly wiped out many local handicraft occupations, and the villagers stopped producing them either for self-use or for trade at the local market. Therefore, the flux of commodities on one hand motivates them to earn cash, and on the other hand increases their dependency on cash.

With a "system of birth-ascribed stratification" (Potter 1983: 465), a Chinese citizen is classified at birth as a rural or urban resident which is then most effectively enforced through the rigid *hu kou* (household registration) system. While in Fenghuang, I frequently heard the local peasants expressing their awareness of such differences in saying "we are peasants, not like you people who eat the state's rice," and "even if you had money, there is little stuff [commodities] available to buy here in our place." Potter (1983: 484–485) discusses possible social mobility through three routes including serving in the armed forces, rising within the organization of the communist party, and making great scholastic achievement. For many ordinary families especially those rural ethnic poor who have no connections or influence, rising up through education seems to be the most feasible means to escape their hereditary social status. In reality, due to poverty, the majority of them could not afford to continue school. In many Miao villages in Fenghuang, the vast majority were illiterate or had only a grade school education.

The nationwide system of valuation attached superior status to urbanites and state-salaried workers, and more recently, white-collar employees in big companies. The lack of education kept almost all rural poor from those higher status positions, and thus, their opportunity to become urbanites. What remains available to them is to work as agriculture laborer, local wage laborer, migrant wage laborer, and maybe self-employed petty business operator, in the relative hierarchy from lower to higher status. Progressive monetization of relations in rural life produced the common perception of valuing capital over labor, and higher-income labor over labor that earned less. The decline in commitment to agriculture and increasing reluctance to undertake it was commonly found in rural China (Croll and Ping 1997), as the marginal returns from agriculture labor were well below those in other activities (Cook 1998). During my interactions with the local peasants in Fenghuang, I often heard such comments as "doing agriculture work makes no money," "men should make *da qian* (big money)," and "only incapable men have to be left to agricultural work in the field."

Zhou et al. (2008: 531) conclude in their survey research of three villages in Sichuan of China that with the shift to market economy, "[the] household that can diversify their income sources away from agriculture, especially to small business, usually achieve higher income levels," and that "as households have shifted from labor to capital as their primary means of production, the basis of income inequality has also shifted from labor to capital." The desire to be self-employed may be understood as an ideological result of the new system of inequality accompanying the penetration of market economy. Along with the spread of modernization and consumerism, rises the aspiration of small-scale entrepreneurial endeavors. It is the self-employment outside agriculture and wage employment that differentiates a person socially, as observed in a Greek village (Galani-Moutafi 1994). A similar phenomenon occurred in Fenghuang. When San Ge was twenty years old, he rented a front shop in the main street of the market town in Shanjiang, and had his own small business as a cotton comber (making quilts) for a while. "At that time people thought that I was a very capable young man," he told me proudly. He believed that it was why his wife, then a good-looking girl who he met there, was attracted to, and soon married him.

In Fenghuang, the booming tourism industry is supposed to provide small-scale business opportunities for the local Miao peasants to make cash, as it requires little or no qualifications. Therefore, it has the potential to enable them to achieve a higher income level to make ends meet with the increasing costs (especially driven by the tourism market) of daily necessities and acquire modern material goods. However, with more and more tourist attractions and land use rights being leased out to outside private developers, the majority of the local ethnic poor were gradually excluded from small businesses with decent profit. Many of them had to drop out of the picture, because they had no access to now privatized tourism resources, because they could not afford the skyrocketing rent for a shop or stand, and because they were marginalized further amid the economic and political forces' construction and reconstruction of local tourism space as discussed in Chapter 3 (Photo 5.1).

Photo 5.1.   A local Miao woman and two boys were selling flower-wreaths to tourists in the heart of the *Gu Cheng*, where nearly all the shop fronts were rented to outside itinerant merchants. (Photo by Xianghong Feng, 2010).

FINDING A PLACE IN THE "PROSPERITY" OF TOURISM

In his writing about China in the 1930s, Tawney (1966:77) describes the position of peasants as "that of a man standing permanently up to the neck in water, so that even a ripple might drown him." This metaphor vividly depicted Chinese peasants' consistent fight for survival back when suffering was a daily ingredient of a life during war and famine. It would not be too much of a stretch to apply it to many of the peasants in today's rural ethnic China, not in the sense of material condition, but in the sense of the hardship they experience in their struggle to "move up." In the following, I document the experiences of a few among those Miao peasants involved in tourism-related entrepreneurial endeavors in Fenghuang, and look at their intentions, desires, and struggles. I assess the choices and constraints of these local lower-class Miao, who represent both common villagers and village cadres, in their attempt to shift the primary source of family income from labor to small-scale capital. The emphasis is on the agency of the Miao peasants who tried to climb out of their different, yet similar, plights, to improve their living conditions with whatever resources they might have been left with.

The Miao communities in rural China are not radical egalitarian (Schein 1997). I present below the Miao villages in Fenghuang as social settings in which daily collaborations and negotiations were developed under the impacts of tourism that aggravated internal inequality and differential distribution of burdens and rewards. I pay attention to how the Miao peasants are internally divided with the intrusion of the external economic forces, and reveal the strengthening of village officials' modest economic gain through their facilitation of the private tourism developers' pursuit of profit. However, it is important to note that my main focus here is not on the political advantages of the Miao village officials, but that despite such advantages for some Miao peasants, they, as those at the bottom of local social hierarchy, all encounter difficulties at various stages in their small-scale entrepreneurial endeavors amid the "prosperity" of the local tourism market.

Choice of a Repatriated Migrant Worker

In his mid-thirties, San Ge had worked as a migrant worker for eight years before he returned to Longcun in 2010 to be by his dying father's bedside. The villagers like San Ge in Longcun did not start out-migration as wage laborers until the early 2000s. According to the village cadres, there had been 60–70 percent of the villagers working as migrant workers, most of whom were adult men with no or a few years of grade school education. This however reduced to around 30 percent with the tourism boom in Fenghuang, especially in Longcun itself as Lu obtained the exclusive rights to run a Miao village tour. Most of the migrated workers, including San Ge, out-migrated for wage-jobs at factories in Hangzhou, the capital of Zhejiang Province in eastern coastal China.

While San Ge and his wife were in Hangzhou, his wife ran away with another man. Since then, his old parents had been caring for his two young children back in the village. San Ge had two older brothers. His *da ge* (oldest brother), Jing, lived a stone's throw away from their parents' house that San Ge inherited and lived in Longcun. His *er ge* (second older brother), Ting, moved to the market town of Shanjiang and made a living as a carpenter. Among the Miao, it was the youngest son's responsibility to live with and care for the parents. With his father passing away and his mother's health deteriorating, San Ge decided to stay. After all, life as a migrant labor in the city was harsh, and he wanted to be able to take care of the old and the young. With several tens of thousands of yuan saved over the years from his wages, he wanted to start a small tourism business.

With the promotion of countryside tours, transporting tourists appeared to be a profitable business. San Ge spent 10,000 yuan on a driving school and received a driver's license. The rest of his savings along with a loan of 20,000 yuan from the local agriculture credit union helped him purchase a new minivan. He had neither the financial capital nor the *guanxi* (social network) to obtain a commercial license. "We thought it would be good business," San Ge asserted. "If it were not because I am so scared of getting caught by the traffic police and paying a heavy fine, I would have had a lot more business." He was caught twice and his minivan was towed to the county Traffic Management Bureau. With a friend's help, he eventually got his van back without turning in the fine. "But it still cost me 200 yuan to take the friend out for dinner," he said. San Ge was not quite sure if he wanted to continue this business, and said, "I am tired of thinking these days. I have to think hard all the time how to make money. It is not like working as migrant labor when I barely needed to think. I knew exactly how much I would end up making every month."

The next day when Lu came to Longcun for business matters, San Ge approached him to ask if he would be interested in hiring him as a driver. Lu offered him a monthly salary of 1,200 yuan with no additional compensation for the gas fee. San Ge thought Lu's offer was a polite gesture of rejection, as it was too low for him to consider. He said, "I could not provide for my family with that salary. I made twice as much salary as a migrant worker in Hangzhou!" San Ge could not hide his disappointment, as it was his last attempt to keep his minivan. He was certain on giving up his van rental business, even before spotting an alternative.

I soon left Longcun for Tuo River Town. A few days later, I received a phone call from San Ge. A man from Tuo River Town was interested in his minivan. He asked him to drive down to Tuo River Town the next afternoon to show him the minivan. San Ge wondered if I might be available to go with him. He was nervous and hoped I, as a city person with good education, could help him negotiate a fair price. The deal was settled at the man's offer of 39,000 yuan. When the buyer drove away in the minivan, San Ge put the 39,000 yuan back in the bank, from which 20,000 was to pay off the loan that he used to purchase the van.

San Ge returned from Hangzhou in September 2010 with 40,000 yuan, which he traded for with eight years of sweat as migrant wage labor away from home; nine months later, he was left with 19,000 yuan and a driver's license but no vehicle to drive. San Ge seemed, however, somewhat relieved after selling his minivan. He no longer had to worry every day about his business, neither would he feel tense while driving. It took courage as a new driver to travel daily between remote villages in mountainous areas with no motor roads. Since he started his van rental business, his mother and his ten-year-old daughter always waited for his return at night by the entrance of the village school where he would park. At least he was happy that he himself and his family did not have to worry about his safety anymore.

Still not sure about what to do next, San Ge wasted no time to explore any possibilities. One of his cousins was interested in doing a tourist guesthouse business with him. They went to look at the rental property in the *Gu Cheng*, and realized they would need at least 200,000 yuan to start the business. The local real estate price increased dramatically with Fenghuang's tourism boom. The monthly rental fee for a front shop had been a couple of hundred yuan ten years ago; now it was tens of thousands. San Ge then went to check out a small rice shop in Tuo River Town. It seemed more realistic than a guesthouse business. But after some simple math, he was not confident to make a minimum profit of twenty cents per *jin*, in order to keep the business running and provide for his family. He was thinking about leaving for Hangzhou again.

Two days after selling his minivan and in the middle of his search for new small business opportunity, San Ge's eight-year-old son fell off the ridge of the un-walled courtyard and hit his head on the quartzite trail below. He took him to a hospital in Tuo River Town after trying the clinic in Shanjiang. When he called me, I was at another hospital with Liang and Hong to pick up their cousin who gave birth to a baby girl there. I left immediately. When I saw San Ge, he was waiting outside the Intensive Care Unit (ICU). He said to me, "I did not know who else to call. You are the only one I know well here in town. Thank you for coming. Maybe you can help me talk to the doctor." With internal bleeding in the brain, his son was in critical condition. The hospital bill cost San Ge what he had and some more borrowed money. He was glad that he had sold the minivan to have the money in hand. With his son in the ICU, his only thought was that he would never out-migrate away from his family ever again.

As his son slowly recovered, San Ge's fear gradually switched from his son's life to his son's future and the future of his family. He not only used up all his savings, but also ended up with new debt. The reality did not seem so pressing when his son's life was in danger, but he now had to face it. He said to me, "I really want to stay close and take care of two young kids. What is the use of money if I lose them? But I have no option, I have to go out again at least for a few more years to make some money. The kids are growing up, and I would like to be able to afford for them to continue school. Otherwise, they would end up like me. I have to go." He made up his mind to repeat his journey as a migrant worker, starting all over again.

Ten days after his son was released from hospital, San Ge left for Hangzhou. Before his leaving, he spent two days building up a brick wall around the courtyard of the house to prevent similar accidents from happening again. He finished as much work as he could in the field to leave as little as possible for his mother later. He then consulted with someone from his village for an auspicious date for his leaving, and invited a shaman from a neighboring village to come conduct a blessing ceremony for his family's health and his moneymaking journey as a wage laborer in Hangzhou.

Struggles of a Peasant Boss

Wen was the developer of "Wanrong River Rafting," a tourism recreational attraction in Fenghuang. He was Hua's uncle. Hua took me to visit Wen in their home village of Lacun. It was in Sangongqiao Township, an almost one-hour bus ride from Tuo River Town toward the direction of Jishou. The bus dropped us at the center of the township. We stood by the road in front of a rice noodle shop, waiting for Wen to go together to the village. Across the street was a branch of the Hunan Rural Credit Union, where Wen was applying for a loan for his business again. While waiting, Hua ran into a teenage boy from the Lacun. He was leaving for Hangzhou to work as a wage laborer, and insisted on treating us to a bowl of rice noodles. After we finished the rice noodles, Wen arrived, with a *hukou bu* (household register booklet) and a *shenfen zheng* (identification card) in his hands. "It's getting hard to borrow money from the bank," Wen said. He frequently visited the bank since he started his river rafting business.

Wen was a Miao in his forties. He had been a migrant worker for fifteen years before his return in 2007. He did not save much from his wage, as it was used to pay off the family's debt, a portion of which resulted from one of his brothers' college education costs. "You couldn't have much economic income by doing agriculture work. The family had counted on me to pay off the debt with my wage as a migrant worker." He claimed all that he had was 3,000 yuan when he returned. With his sister-in-law as the guarantor, he pledged for a bank loan, and started a brick factory with a modest investment of about 30,000–40,000 yuan.

In 2008, numerous factories closed due to the bird flu and the global financial crisis. Many migrant workers returned to their villages including Lacun. It was around when Fenghuang's tourism was promoted by the county government from Tuo River Town to the countryside. Wen and his returning fellow villagers decided to run a tourism-related business. "We came together because we were in a similar situation. There was not much for us to do here, and no place to make money," Wen said. At that time, there was one river rafting company called "Ximen Gorge Rafting" in the upper part of the same river running through Lacun. Wen recalled how he was inspired, "It is the same river, similar scenery. If they could do a good business, so could we."

They estimated an investment of 300,000 yuan. As there were sixty of them, each of them needed to put in 5,000 yuan. For those who could not afford it, they invested with their land and wood. They did almost everything themselves to construct a river rafting route, except the explosive projects. They soon ran into problems. As the rafting route ran through two neighboring townships besides their own, they had to rent some land. Through negotiation, some of those villagers were willing to lease out their land at a price of 6,000 per year, while others were more interested in joining the business. Wen said, "There were quite a few of them wanting to join. Why? Because it is so hard to make money here."

After Wen and his fellow villagers spent nearly 50,000 yuan to clear half of the rafting route, and almost 200,000 yuan to build a dam, there was not much of their pooled money left. Many of them quit, since the demand of further investment far exceeded their original estimate. Only fifteen of them stayed. Wen understood the situation for those who quit, and said, "It is so hard for people like us to make money. If they had to put in more, they normally would not have the guts. Neither would they have the insight. They were not sure if it would turn out to be *da shuipiao* [playing ducks and drake, meaning throwing away money] or else. As I was the one in charge, I ran around to find more money to invest."

With one year of hard work, 2,000,000 yuan, and six of them, they opened "Wanrong River Rafting" in 2009. Wen ended up with a bank debt of 300,000 yuan. As each person could only get a small loan of 30,000, Wen asked all his relatives and friends to apply for loans for him in their names. Wen said,

"I tell you what, now in the entire village, I have bugged every one of them whom I have an okay relationship with to apply for a loan for me. I now have a whole bag of personal seals I have had made." Among the five besides Wen who stayed at the end, three were Wen's brothers, and the other two were brothers from another village, also relatives of Wen. They had a clear division of labor. Wen's older brother, the one who went to college, was an official with the county Forestry Bureau, and he took care of social networking at the upper level; one of Wen's younger brothers was stationed in their reception center in the *Gu Cheng*, in charge of marketing and receiving tourists; the responsibilities of the two brothers from another village focused on safety supervision and personnel management; and Wen oversaw everything, including solving all major problems and attending meetings organized by the county Tourism Bureau.

They bought 30 boats at prices ranging from 5,000 to 8,000 yuan, and hired more than 30 people, the majority of whom worked as boatmen. Many of them were relatives. "People told me not to hire relatives," Wen said to me. "But you see, it is always difficult in our countryside. For example, my nephews and nieces are growing up and have nothing to do. If I did not let them do some things here [in my business], what else can they do [to make a living]? Leave them at home playing every day? That is impossible [for me to do]. It was best if we could solve our own problems. This way I could partially help with the problem of the surplus labor for our own relatives, villagers, and local people. Right? That was what I thought then."

It was hard to establish a business, and even harder to run it. Wen's major concern included personnel management and marketing. His biggest concern was to deal with some of the employees who he hired from the other two townships. It was not an easy situation for Wen as the boss when part of his business depended on the access to his employees' land. If they were reluctant to do something assigned by Wen, he had no good way to make them. He wished that the rafting route only involved the land in his own village. "I don't have managment experience, nor the brain for it," Wen said. "We need someone who is good at it, whom however we couldn't attract to work for us. It is simple. I have to pay a decent salary for the qualification, but I don't have that money." In addition, the business was seasonal, making it even harder for good hires.

Wen's business received 3,000 tourists during the open season from April to September in 2009, and about 10,000 in 2010. The gross income was far from enough to cover the costs. Wen was concerned with the marketing. As a new business, he cared more about its *ren qi* (popularity) than *zheng qian* (making money). Wen tried multiple strategies to draw tourists to improve its *ren qi*. First and foremost was to lower the ticket price. The printed ticket for "Wanrong River Rafting" showed an "official" price of 148 yuan, but Wen set the actual *di jia* (bottom price) at 78 yuan. Thus, it left a decent profit margin for ticket dealers (e.g., travel agencies, tour guides, and tourist guesthouses), which served as an incentive for them to take more tourists to his business. Wen was also initiating a partnership with one village to bundle his river rafting with a Miao village tour. Wen was not quite sure if he did the right things, but was hoping more people would know "Wanrong River Rafting" soon.

The "Ximen Gorge Rafting" was his biggest business rival, developed in the early 2000s by people from Tuo River Town who had been associated with the Fenghuang Tobacco Factory. In Wen's eyes, they had various advantages. He said, "[They are] different from us. We are peasants, and they are from *dan wei* [work unit]. They have better networks, stronger relations with the government, plus their money, their brains, and their *wenhua* [literally means culture]. Whatever aspect, you name it, they are better, and we have almost nothing to compete with them. The only advantage we might have is that we are at our own place, at least part of it, doing business, while the land that their business is on is all rented." The key to win the competition lies in the marketing, in which sense, for Wen, it was not much different than fighting a war, like the one between the Communist Party (at its infant stage) and the Guomindang in the early 20th century.

At the starting point of the rafting route, Wen constructed vending stands to sell snacks, drinks, raincoats, and other small goods. His original plan was to have them rented out. As the rafting was seasonal, and the tourist numbers not yet stable, there had been little interest. Wen then required that each of the shareholders send someone from their individual household to take one stand. "No matter if there were many tourists or none, we must operate these stands; otherwise, it doesn't look good as a tourism attraction lacking of *ren qi*." He was firm on this, saying, "Every shareholder must do one stand. I set myself as an example, and sent my wife there." Wen joked that his wife was richer than he was, "She does it all right. She pays for the kids going to school. Any little bit of money is from her. I have no income but keep on losing money. As you see today, I was applying for more loans earlier."

Wen's confidence started fading away with the increasing difficulty of running the business. He was not sure when he could have at least his investment back. He wished he could sell it to the county government or some private investor. I noticed there were some bonsai in his courtyard. He had already begun to learn making bonsai. He thought it might be a promising business. As real estate had begun to boom in Fenghuang, he was positive about the demand. But for now, he had to try to take care of the rafting business. He decided to make marketing efforts himself outside Fenghuang. He was uncertain about the turnout, as he said, "This is very difficult. For us who don't have *wenhua*, this is extremely difficult. Am I right? We don't know how to say a lot of things in Mandarin, as we are Miao. We don't know computers either." Nevertheless, once he received the bank loan he just applied for, he would immediately leave for Changsha and Chongqing, the two main cities of origin for Fenghuang's tourists.

Dual Positions for Village Cadres

Qiu, the village Communist Party Secretary in Longcun, had recently acquired another position. Lu, while not interested in hiring San Ge as a driver, offered Qiu a managerial-level position in his tourism company, with solving disputes between Lu and the villagers as his main responsibility. Qiu's family was among the few households that were benefiting most from Lu's tourism business in Longcun. Qiu's family restaurant catered meals to tour groups. Opened in late 2009, it was the first one in the village. He lost 500 yuan during the first month's business. It was not the peak season, and he overestimated the number of tourists. His family ended up eating most of the vegetables and meat. The second month, he lost 250 yuan, and the third month, he made a profit of 300 yuan. Two years later, his average monthly profit increased to above 1,000 yuan. He said, "To be honest, we are now starting to have passbooks. In our village, no one had savings in the bank before, only loans."

Encouraged by his success of the restaurant business, Qiu and his co-investor (someone from another village who also worked in Lu's company) rented a property in the *Gu Cheng* to run a tourist guesthouse in early 2011. They together invested more than 300,000 yuan, including 200,000 yuan as the first two years' rent, a business transfer fee of more than 70,000 yuan, and a deposit of 50,000 yuan. As Qiu's dual duties as a village cadre and a manager at Lu's company kept him busy, he had his sister-in-law be in charge of the guesthouse. For the first three months, there was not a single guest. "The business was not that good," Qiu said. "It is harder for people like us to do business in Tuo River Town, who neither traveled far nor did business before. We couldn't compete with others, as our economic brain can't match theirs. . . . It is like gambling for us. If we two did not manage the guesthouse well, I would lose my shirt, and my restaurant business would be ruined as well."

Qiu had three sons. The youngest one was ten years old. The oldest one was 23 years old, and out-migrated a few years earlier. Qiu wanted his second son, who was 20 years old, to stay and learn to run the guesthouse. One day, Qiu received a cell phone text message from him, saying, "I am on the train now. Don't come and chase me back." It turned out he contacted his migrant cousin who sent him 3,000 yuan for travel. Then he left Fenghuang surreptitiously. When I asked Qiu why he did not want his second son to be a migrant worker, he said, "I want to keep him home to get some experience of doing small business." "But he was not interested?" I asked again. "He told me when he saw no guests for two months, he was panicked. He thought that it cost me so much to invest in it, so he wanted to go out to make some money for me," Qiu replied. Despite the initial hardship, Qiu's networks with Lu and Lu's other employees (especially the tour guides) did help him with his small-scale entrepreneurial endeavor, more so for his restaurant than for his guesthouse business.

Like Qiu, Ping, the village Communist Party Secretary in Wucun, also worked for the tourism developer of his village, Ma. Ma and Ping had a close personal relationship, since Ping was the one who carried Ma on his back home when Ma fell of the cliff and broke his leg. As Ma spent more time in Tuo River Town for marketing and networking, Ping was the one to operate the tour business on a daily basis in the village. In Wucun, even though the conflict between Ma and the other villagers was not as tense as in Longcun, Ma relied on Ping for coordinating things between him as his fellow villagers as well as writing reports and marketing materials. Ping said, "To tell the truth, [Ma's business] was counting on me for all its software." Ping had been the village accountant for three years, thereafter holding the Party Secretary position for six years. With that experience and a middle school education, he was one of the few who had the most *wenhua* and people-management skills, and thus was an asset to Ma for his business. One of the main duties for the village cadres like Qiu and Ping was to adjust matters and harmonize clashing interests between the developers and the villagers.

A strategic comradeship ties tourism developers with local political authorities. This business alliance between private developer and political authorities did not only exist at the village level. The higher level the position was, the better was the potential return. One of the six shareholders of "Wanrong River Rafting" was the brother of Wen who was an official at the county Forestry Bureau. He had a good relationship with Jun, who was the deputy general manager of the Xiangxi Fenghuang Holiday Travel Agency as mentioned in Chapter 1. In Jun's travel agency, the brochures and posters of "Wanrong River Rafting" were placed and hung in the most obvious spots; and Jun's office staff and tour guides always recommended it to the tourists. Jun frequently offered Wen advice on management and marketing based on his over ten years experience running a travel agency. Jun also accommodated Wen's niece Hua as an intern in his company. The social networks of Wen's brother helped much with Wen's (as well as Jun's) business. It likely contributed to Wen's relative success, despite hardship, in comparison with other Miao peasants such as San Ge.

TOURISM-RELATED "POVERTY OF RESOURCES"

With the precapitalist agrarian society being increasingly drawn into the market economy, rural livelihood becomes significantly dependent on cash earnings. The surplus labor, diminishing return of agricultural cultivation, and rising costs of daily necessities are all "push" factors for peasants' leaving their field (Croll and Ping 1997; Song 1996). Even amid the economic "prosperity" brought by tourism in Fenghuang, many poor Miao peasants like San Ge seemed to have a hard time finding a place in this "prosperity." After his failed attempt of doing a van rental business and his son's accident, San Ge said, "I am so poor, so poor that I am now worried about having enough food to eat and clothes to wear." Later, when San Ge set off for Hangzhou again, he was joined by two fellow villagers, who quit working for Lu as boatmen earning 600–700 yuan per month. "It is a good thing that both of them are unmarried. They could only afford feeding themselves with that salary," San Ge commented. He hoped to return with enough savings, besides covering for his children's school fees, for maybe another try in small business.

Due to the capital constraints, the poor peasants have to rely on their own funds to operate new enterprises (Chang 1993; Ho 1995; Mohapatra et al. 2006). Labor is the poor's greatest asset (Cook 1998; Moser 1996, cf. Gonzalez de la Rocha 2001: 81). Off-farm labor earnings were the main source to accumulate such funds (Rozelle et al. 1999; Taylor et al. 2003). In rural Fenghuang, it was typically the wages earned as migrant workers in cities. San Ge had no choice but to repeat his out-migration journey in order to provide for his family and maybe have another chance in self-employed small enterprise. The continuous incoming of such funds was also important to sustain their business at least during its early stage. Qiu's second son, against his father's will, ran away from home to be a migrant wage laborer, wanting to make up the money his father had lost in his newly opened guesthouse. Even though somewhat more successful than San Ge, Wen was ready to give up his river rafting business once there was a way out.

Despite its prominence in the literature (Mohapatra et al. 2006), the capital constraints, however, may not be the key barrier that kept the rural poor from succeeding in their small-scale entrepreneurial endeavors, considering the availability of small loans from the local agricultural credit union and their own savings from wages. What might be more crucial was the sociopolitical means and will to ensure and protect their access to local resources, which had long been rightfully theirs but were now gradually monopolized by private investors in the development process, therefore creating the growing "poverty of resources."

More than a half century ago, Fei and Chang pointed out, "Given the opportunity, China will inevitably be industrialized; but whether or not this new industrialization will be beneficial to the peasants is the problem" (1945: 308). To improve the livelihood of the Chinese peasants, Fei (1939; Fei and Chang 1945) repeatedly emphasized the necessity of increasing their income through decentralized rural industry, and the importance of the profits brought by the rural industrialization being distributed as widely as possible, as he claimed that "the effect of such industrial development . . . is bad, so far as the poor villagers who have no opportunity to participate are concerned" (Fei and Chang 1945: 310).

In Fenghuang, new forms of exclusion of the poor came with the elite-directed tourism development, which provided resources to those already rich and/or with political advantages. The support from village cadres and local officials could be crucial to the success of tourism developers' businesses. In exchange for assistance derived from their political offices, the developers granted them access to economic benefits, from which most peasants were generally excluded. What lies behind the rational decisions of these peasants regarding what to do to make a living and how to best provide for the family mirrors the social realities they, as those who are at the bottom, are dealing with. Their dilemma is that they have to struggle to live off the land and seek security beyond agriculture, while the land is the only real source of security.

The façade of prosperity can often mask the nearby precipice (Beck and Beck-Gernscheim 2002: 3). Under the shadow of this tourism "prosperity" were the vast majority of Miao peasants, who were illiterate or nearly so, spoke no or some Mandarin (but not well), and rarely had many connections to the upper social class. Village cadres and their families might have enjoyed better opportunities from economic growth (Cook 1998; Walder and Zhao 2006). However, despite the limited advantage derived from their relatively low political office (in the case of Qiu and Ping) or their kinship tie to such position (in the case of Wen), they, the lower-class Miao peasants like San Ge, all experienced hardships in their small-scale entrepreneurial endeavors.

The critical issue here, I emphasize, is the access to resources. The only advantage mentioned by Wen in his business was that his village maintained the use right of part of the Wanrong River. Had it been privatized to outside developers, Wen and his fellow villagers would have had no resource to develop their river rafting business, even with plentiful labor and funds and Wen's kinship ties to a county official. In contrast, in Longcun with its village tour monopolized by the outside developer Lu, San Ge had no access to the tourists in his own village, who were potential customers for his rental van business, and he had to commute to Tuo River Town to seek tourists; and Qiu's relative success with his restaurant business depended on his access to these tourists in Longcun, which was allowed by Lu in exchange for the advantages derived from his political position. The current living condition of local Miao peasants is better described as the "poverty of resources"—the lack of equal economic opportunities and the erosion of social protection in a context shaped by an exclusive tourism development model. In Table 5.2, I summarize the resources

Table 5.2. Common ways of making a living among the Chinese peasants.

Resources of Poverty	Land, water, and other natural resources support subsistence	Petty commodity and trade production provide additional income	The informal social guarantees of village life (e.g. kinship network and reciprocity)
*Poverty of Resources**	Natural resources (land, water, etc.) as well as historic and cultural resources are either privatized as tourism attractions, or lost for tourism related infrastructure and real estate development	The intrusion of manufactured commodities; as the return diminishes, the petty commodity (including traditional ethnic handicrafts—an important tourism resource) production disappears	A decline of earlier social insurance due to the progressive monetization of relations in village life; the out-migration of adult males weakens mutual help between village households as the Miao are patrilineal and patrilocal

* Note that the "poverty of resources" already took place in Fenghuang prior to its tourism development through the penetration of market economy. Here I mainly highlight those changes directly associated with tourism.

that had been crucial for the Miao peasants' survival (the resources of poverty), and juxtapose them with the erosion of these resources (the poverty of resources) resulting from tourism development (Table 5.2).

Khan and Riskin (2005) state that rural non-farm activities have been a highly disequalizing income source, and suggest promoting rural non-farm activities while improving the access of the poor to these activities to increase income. However, as Naughton (2007: 14, cf. Zhou et al. 2008: 515) states, "Government policy has been focused on market liberalization, social protections have been ineffective and eroded, and an unfair and unequal market economy has emerged in China." With the commercialization of rural economy, the increasing dominance of the market as a mechanism for resource allocation inevitably creates the "poverty of resources" for the peasants, the majority of Chinese society. China's current rural development policies in ethnic areas have been largely relying on the privatization of local natural and cultural resources, in which tourism development is a common strategy. They are designed (though unintentionally) to fail to improve the living conditions of the poor, as they decrease the amount of resources in their hands. In an exclusive economy monopolized by the elites, only the wealthy get wealthier. The key to bringing about equitable growth depends on a more moral, rather than a solely market, allocation of resources.

NOTE

1. This chapter is revised from "From Labor to Capital: Tourism and the Poverty of Resources in Rural Ethnic China," which appeared in *Urban Anthropology and Studies of Cultural Systems and World Economic Development* 41 (2, 3, 4): 329–365, Fall 2012.

Part III

ONE VILLAGE

Chapter Six

Before and After the Merger

Everyday Resistance in Village Life

It was one summer afternoon in Longcun. Right next to the village front gate was the performance area, composed of a cluster of structures, including a central stage, a circular ground in front of it, two seating areas flanking it, a row of vending stalls around it, and a public restroom next to it for tourist use. The speakers were silent, and the seats were empty. No shows, and no tourist crowds. Four children were running around on the stage, playing with the drums. Two girls in their work outfits (the simplified Miao costume) sat by the gate, chatting. In early 2012, in the name of better regulating the countryside tour market, the county government merged eighteen countryside tour sites, including Longcun, and placed them under the management of its newly established Fenghuang Folk Culture and Custom Tourism Development Co. Ltd. (FFCCTD). The original developer of each site was paid with the combination of cash and certain shares in the FFCCTD. Many of them took managerial-level positions at the FFCCTD, including Lu, who became the head of its Construction Department. The FFCCTD shut down ten sites and developed several tour routes with the remaining eight. One route featured Longcun and the Shanjiang Miao Museum as a one-day tour.

This was a typical scene in Longcun since the merger (Photo 6.1). The two girls were among the eight employed by the FFCCTD to conduct the "three roadblocks" welcoming ceremony. They came to the gate at 7:30 a.m., had breakfast, got ready for the arrival of the first tour group, and continued until 5:00 p.m. The monthly salary was 1,000 yuan. They worked on weekends, with four weekdays off each month. One girl complained that their job was much more demanding than the performance troupe, as the troupe only need do shows at most three times a day, while they had to sing, drum, and toast

Photo 6.1. A village dog was resting on the central stage of the performance area, with a few tourists wandering around and taking pictures. (Photo by Xianghong Feng, 2013).

numerous times whenever tourists arrived. One big wooden board hung by the gate listed the names, positions, and headshots of the villagers who were the FFCCTD's employees: eight welcoming ceremony receptionists, ten show performers, four guards, four sweepers, and three logistics personnel. The performers, however, were already dismissed upon the cancelation of the show two months into the merger.

EVERYDAY RESISTANCE

Revolving around Longcun both before and after the merger, this chapter spotlights everyday resistance, the informal and often individual acts, through which local Miao peasants attempted to fight for a share of benefits in the current economic context of elite-directed and capital-intensive tourism development. From a ground level, it focuses on how Longcun villagers exercised their power, no matter how limited it might be, to resist not tourism itself, but the uneven distribution of benefits from tourism.

James Scott's *Weapons of the Weak: Everyday Forms of Peasant Resistance* (1985) led the search for less organized, more pervasive, and more everyday forms of resistance. Rather than collective action with sustained protest efforts, everyday resistance is typically individual based, and normally aims not at grand reforms but specific demands (Adas 1986: 69). It "involves intentionally contesting claims by people in superordinate positions or intentionally advancing claims at odds with what superiors want" (Kerkvliet 2009: 233). Among the existing scholarship of resistance studies on China (e.g., Li and O'Brien 1996; O'Brien and Li 2006; Walker 2006, 2008; Zweig 1989),[1] few look at non-policy-based everyday resistance (especially that by the ethnic minorities) from a ground level.

Resistance studies grew rapidly over the past forty years (Fletcher 2001; Ho 2011), so did the criticism of them.[2] Among these various arguments, one major line of criticism was the lack of "thick" ethnography (Geertz 1973) to profile peasants as the resisters, portraying their culture, internal conflicts, and individual motivations (e.g., Fletcher 2001; Ho 2011; Ortner 1995; Seymour 2006; White 1986). As Ortner (1995: 173) stated, "Many of the most influential studies of resistance are severely limited by the lack of an ethnographic perspective." She described resistance studies as ethnographically "thin" on the internal politics, cultural richness, and subjectivities (e.g., the intentions, desires, fear) of subordinated groups (1995: 190).

Scholars have taken steps toward ethnographic "thickness," to enable the ethnographic subjects to reveal their social world in ways that make sense to them (Brown 1996: 733). Such efforts have been applied to the studies of Hmong resistance in the uplands of Vietnam and Thailand (e.g., Michaud 1997, 2012; Turner 2012a, 2012b; Turner and Michaud 2009).[3] Despite these efforts, there remains a need to further illustrate how particular resistance incidents configure among a certain group of peasants, specifically focusing on the peasants involved in these activities and their reality in terms of who they are, how they live, what they think, and the historical particularities of both the group and the local and national situation (Brandtstädter 2006; O'Brien 2013).

The following discussion pays particular attention to the complexity of internal village politics induced by tourism. It undertakes deeper analysis of these peasant resisters' dynamic interpersonal experiences of constraints and resources, to understand how they themselves viewed their situation, why they had reacted as they had, and what factors accounted for their adoption of different forms of resistance. It explores how the structure of tourism-power relations determines the effectiveness or futility of everyday peasant resistance, and concerns if and how everyday peasant resistance narrowed the range of options available for the dominant regarding tourism development. It argues that these informal and individual activities, rather than the contentious collective political activism, are the more typical *everyday* form of peasant resistance in contemporary China.

LONGCUN AFTER THE MERGER

Qiu described the merger as *daguo fan*, the communal-dining system during China's Great Leap Forward. To the FFCCTD, as long as the overall number of tourists was good, it didn't matter which site(s) these tourists chose to visit. Wucun was also merged into the FFCCTD, and its developer Ma, just like Lu, became a shareholder of the FFCCTD and was in charge of maintaining stability. When talking with Qiu about the merger, I mentioned Ma. "Ma is our relative. He was from another village here in Shanjiang before he moved to Wucun," Qiu said. "Just the other day I met him in Tuo River Town, and he took me out for a meal. He said frankly that no matter what, the merger was better for him. He was certain that otherwise Wucun's tourism would die sooner or later."

I visited with Pang Ge in a café in Tuo River Town later that summer. Pang Ge told me that Ma had much more leisure time since the merger, and they drove around to go fishing now and then. I remembered my last visit with Ma at the teahouse. It was about half a year before the merger, when Ma was so concerned about Wucun's future that he was looking for investment to start a Peasant Family Happiness project (as mentioned in Chapter 1). With the money from selling the tourism rights to Wucun, Ma was ready to get this project in full swing. The merger was indeed better for Ma. As Ma himself put it, "I have money coming in even if I sit at home. No worries for my own salary, and no worries for the salaries of my employees. Who cares how many tourists visit Wucun?"

However, to the villagers at each site, the number of tourists was all that mattered. As the villagers perceived, Longcun's tourism was "worse and worse." Mi, the village treasurer for over ten years, said, "From what I have seen, tourism in our village is going downhill this year. This perhaps is the common impression among us [the villagers]." He continued, "Lu does not come, neither does the government. It seems that we are left on our own." Some of the problems during Lu's management became worse under the FFCCTD. Mi complained that those working for the FFCCTD in Longcun played poker all the time at the village parking lot. "They get paid for sitting there being lazy all day." Mi said, "If the FFCCTD does not care, we [the village cadres] could do nothing about them. But the FFCCTD has kept turning to us for assistance."

The FFCCTD appointed an on-site supervisor in Longcun. Whatever issues the supervisor couldn't handle, he called Qiu for help. Qiu was annoyed, and finally told him, "Don't always count on us to help you solve problems. You are the supervisor, and you are paid to do the job. It is not like we [the village cadres] are not willing to support your work. You have to rely on yourself. We can only assist. We can't take the lead. We are not the ones who get paychecks from the FFCCTD anyway." Qiu told me that his willingness to assist the supervisor was for the sake of the village, not for the FFCCTD. He said, "If there is something not handled well, we are afraid that it may drive tourists away." Words circulated that many government officials and developers arranged for their relatives to work at the FFCCTD. Qiu said, "Knowing this, my heart went cold. I help, so that their relatives get paid without doing the work?!" When Jun's wife visited the FFCCTD for business related to their travel agency, she was upset about the staff's arrogant attitude, saying, "To be honest, those people did not know anything about tourism at all." One of the FFCCTD's salesmen was an acquaintance of hers, and told her that many of the FFCCTD's office staff obtained their well-paid positions through *guanxi*.

Despite of the county government's claim that the merger was an official act, some believed that the merger was initiated and sought after by a private developer but was executed by the county government. "Many people had a similar suspicion. The funds were from the developer, but went through the government's hands to smooth out the transition," Hua said. Qiu had believed the merger was an official act until he interacted with the FFCCTD on two occasions. It left him with the impression that the FFCCTD did not seem to have forceful power as the government typically possessed. Qiu asked Lu several times if the merger was the government's idea. Lu responded with "yes." Qiu claimed that he sensed Lu's dishonesty in his unusual low voice. "There must be some big businessman behind the merger, who paid the government to carry it out." Qiu said, "This is my guess. If it were indeed governmental buyout, it should not have been managed this way." I once rode with San Ge's friend from Longcun to Tuo River Town. He was from the market town of Shanjiang, and was employed by the FFCCTD and worked at Longcun's ticket checking point. "The company only exists in name," he said, "and the PATT is in charge of everything, and our paycheck is issued by the PATT."

Along with the merger came frictions inside the FFCCTD. Its upper level was largely composed of the original developers from individual tour sites. They did not get along. "The problems within the FFCCTD are complicated. We have not the slightest clue of exactly what is going on up there," Mi said. Mi and other village cadres visited Lu in Tuo River Town recently. Since the merger, Lu moved from Shanjiang to Tuo River Town. Lu called Qiu several days earlier, and said that he felt sorry for the Longcun people. Lu explained that if he came to visit, his heart would be saddened, so it was simply easier for him not to. Qiu recalled that it had been awhile since Lu's last visit to Longcun. Lu used to be seen in Longcun all the time. I frequently encountered him in the summer of 2011. My first conversation with him took place after our lunch at Jing's house. Lu came to the village that day, with three office staff from his tourism company in Shanjiang. They were waiting for an electrician to work on the lighting system in the cave on the tourist-hiking route to the village's front gate. Back then, Longcun's tourism was his own business, and he had his heart set on it. Indeed, that summer in 2012 I had not seen Lu at all, until my trip with Jing to the FFCCTD.

Summer weekends were always busy for Jing. It was one Sunday, and Jing's restaurant had fed over 400 tourists. It was past ten o'clock at night, and Jing's wife and kids were washing up for bed. Jing had one last thing to do before ending his day. He took out a calculator, a pen, and a small notebook. On the notebook were various marks he put down throughout the day. Those marks only made sense to him. As a school dropout before finishing the first grade, his daily bookkeeping was made possible by the calculator. With the sound turned on, each number he entered was read out loud. Jing sorted out the meal cards, hundreds of them, and bundled them in stacks with rubber bands. Since the merger, the FFCCTD issued these meal cards. The FFCCTD charged the tourists tour fees including meal costs, and distributed the meal cards to the tour guides. The tour guides used the cards to pay the restaurants to which they took their tour groups, one card per tourist per meal. The restaurants later visited the FFCCTD's accounting office to exchange the cards for cash, each worth ten yuan.

"I am going to cash out these meal cards tomorrow morning. You want to go with me?" Jing turned to me, and said, "They will want my signature. I don't know how to write my name, and you can help." Six o'clock the next morning, I was wakened by noises. Jing and his wife were up. The attic of Jing's house was a shared bedroom with two beds, each covered by a mosquito net. The bedroom had once been located on the main floor until it was used as the dining area for the family restaurant. Half an hour later, Jing, San Ge, and I were heading to Tuo River Town. After completing the business at the FFCCTD, Jing and San Ge hurried to the wholesale agricultural market to replenish food supplies for the restaurant. They had to make it back to the village before the first group of tourist customers arrived at the restaurant. I decided to go find Lu.

ONE MORNING WITH LU

"Two tons of sands, two tons of cement—that is 1,000 yuan already. . . . The total would be 6,000 yuan—he must be insane!" Lu was on his cell phone, when I found him. It was in the lobby of a hotel in the same courtyard of the FFCCTD. A contractor at one of the FFCCTD's tour sites had called Lu to obtain purchase approval for construction materials. Once in a while, people from the FFCCTD's tour sites called him or sought him in person for other construction-related issues. In between interruptions, we talked about the merger, officially known as *lian ying* (joint-management).

"It is not exactly *lian ying* [joint-management]. It is more accurate to call it *he bing* [merger], since the government bought out 51 percent of its stock share." Lu explained to me. Lu considered the FFCCTD as an affiliate of the county government. The county government appointed the director of its Legislative Affairs Office as the general manager of the FFCCTD. According to Lu, the county government spent between 20 and 30 million to buy out individual private businesses at each tour site, and gave each of them a certain percentage of the FFCCTD's stock share. As a trustee of the FFCCTD, they also received a fixed return—15 percent of the annual dividend. The county government promised to invest 150 million yuan over 15 years to improve Fenghuang's countryside tours.

The provincial tourism bureau called Lu, requesting data on Fenghuang's countryside tours for research. Lu called Tang, a colleague in the FFCCTD, and asked him to have someone to take care of it. They talked a bit more over the phone. Lu mentioned to Tang that someone else gave him a cold shoulder when he greeted him the day before. They had a disagreement at a board meeting regarding the purchase of a company car. Lu said to Tang, "Let him be! I said what I had to say. We should spend on tour site development, not luxury cars and fancy hotel rooms. Look at our tourism product, and look at our management. It is worse one year after another, one month after another, and you buy expensive cars right now?! . . . If you want to buy a Jetta, I am okay with it, but not a car costing several hundred thousand yuan?! . . . Anyway, I have to go. Don't forget to take care of the tourism bureau thing."

Lu hung up. He complained to me about the friction within the FFCCTD, driven by the board members' personal interests. Lu thought it was a good idea for the county government to consolidate the countryside tour sites, to regulate the "messy" tour market. But he believed it was not wise for the government to compose the FFCCTD's management team with the original developers of individual sites. "Business management is much more than a bunch of peasants putting out several *wan* [10,000 yuan] to set up a Miao village tour site," Lu said. "There is absolutely no [professional] business management here!" I asked, "What specific things are you in charge of at the FFCCTD?" "Nothing!" Lu replied. "Well, I don't dare to take charge, and frankly, I am not in the mood to bother anyway."

Lu's focus was elsewhere. Upon the merger, Lu closed his company in Shanjiang, and was starting his new travel agency business. He sold his partial shares of the FFCCTD for 1.5 million yuan. He was waiting for the transaction to be final, so he could use the money to register his new company. With Lu in the lobby was the secretary he hired for his new company, a girl in her early twenties. She was busy running errands for Lu. Lu asked her to make a reservation for a private room at a restaurant, "Find out if they have good red wine, and how much." He was arranging a business dinner, and this was for his own company. He then sent her to get rice noodles for him as breakfast.

Over his rice noodles, Lu continued, "To the share holders like me, it is a gamble. We don't care. See, we were struggling to run these tour sites ourselves. I gave it to you [the county government], and you [the county government] gave me the money. So I don't care anymore. Life or death, it has nothing to do with me." I asked, "What is your monthly salary at the FFCCTD?" "Twenty-six hundred yuan," Lu replied, and said, "Let's have some tea." He told the lobby attendant, "Use my own tea leaves." He then sent his secretary to buy three bottled waters, saying, "The tap water they use to make tea is not good. It will ruin the taste." I joked with him, saying, "Perhaps your salary is only enough to buy you a half *jin* of tea leaves." Lu laughed, and said, "Not even for very good ones."

I asked Lu when he had last visited Longcun. "Three months ago," he said, then asked, "Longcun is not as good as last year, right?" I mentioned the conflicts between the village and the FFCCTD. Lu commented, "If the government always makes accommodations to the people, it will turn them into *diao min* [shrewd and unyielding people]. It is like how the father should treat the son: if the father never beat the son, the son will never listen to the father, and will never be a dutiful son." He then said, "All the conflicts are between the village and the FFCCTD, nothing to do with me anymore. When looking back, the villagers must be regretful."

Lu was right. Despite the problems the villagers used to have with Lu, they did miss the days before the merger. After the FFCCTD took over, it granted limited authority to its on-site supervisor at Longun. "Its management system is not good," Qiu said. "If anything urgent happens, the supervisor has to go through one level up to another for approval before making any expenditures. By then, it could be way too late." Once a village child threw a rock and injured a tourist's arm. Qiu said to the supervisor, "Take the tourist to the hospital first. If the FFCCTD refuses to reimburse the medical bill, I, as a village cadre, will pay for it out of my own pocket." Qiu told me when he worked for Lu, Lu gave him the authority to make emergency decisions. Qiu also had the liberty to purchase gifts for his visits to other villagers, to take them out to restaurants if running into them in Shanjiang, and to make travel arrangements to Tuo River Town for business. Reimbursement was never an issue.

When the Longcun tour was Lu's own business, he invested in cultivating relationships with the villagers and in improving tourists' experiences. Lu was much more involved in the village business than the FFCCTD. Whenever there was a wedding or funeral in the village, Lu either attended himself or more often sent a representative. "What mattered to us was not the amount of cash gift in his red envelopes, but the respect from him," Qiu explained to me. Lu always consulted with the village cadres for ideas to gain support, while the FFCCTD showed no interest in doing so. If its collective funds were short for a village project, the village could normally count on Lu's contribution. As Lu monopolized the access to tourists in Longcun, the village cadres relied on Lu to mitigate internal competition. For instance, Lu used to distribute tour groups led by his tour guides evenly to the six family restaurants on a daily basis, and he could stop allocating tour groups to a restaurant if it lowered its meal price. After the merger, Qiu requested the FFCCTD to do the same. The FFCCTD responded with "no such energy to do so."

NON-POLICY-BASED "RIGHTFUL" RESISTANCE[4]

However, the fact was that before the merger, frequent conflicts happened between the villagers and Lu. A villager one night sabotaged the drums that Lu had purchased and placed on the performance stage for the show; another damaged the water pipe to the tourist restroom that Lu built by the performance ground for the convenience of tourists; more provoked daily quarrels with tourists or Lu's employees, such as his tour guides, whom Lu recruited elsewhere. Among the daily episodes of conflicts, a series of entangled incidents happened surrounding the vending stall, tourist restroom, and roadblock.

One villager, Min, worked as a parking lot attendant for Lu and had previously obtained from Lu one of the old vending stalls in the performance area for his wife to use. Another villager, who had rented his land to Lu to construct the performance area, was dissatisfied with the location of one of the new vending stalls assigned to him. He decided to instead take the stall used by Min's wife. He argued that even though he had rented the land to Lu, the land was still his. Min felt that he had no grounds to fight back when the original land user claimed it. Neither was he willing to let it go. He made his position clear. He blocked the entrance road with rocks, preventing tour vehicles from entering the village. The township officials and police came to persuade him. At that time, Min's wife was at home under the customary monthlong postnatal care (*zuo yuezi*). One of the police promised him that if he stopped blocking the road, a job with Lu's company would be arranged for his wife after her postnatal care. Min agreed.

Certainly, Min's wife came to Lu's company one day, ready to work, only to find out that Lu had not agreed to hire her. Lu insisted that it was the police, not him, who offered it. He said that the roadblock and employment were two separate issues, and was unhappy with the police's irresponsible promise. Lu was concerned that if he accommodated this, whoever wanted a job from him in the future would be guaranteed one by blocking the road as extortion. "How could I possibly afford that?!" he said. Min was upset. He went to guard the public restroom, charging tourists for using it. The village cadres tried to reason with him, claiming that no public restroom was charged for use in any tourist destinations around China. Min refused to listen.

Despite the fact that Lu had constructed the restroom, the land was rented from another villager, Teng. Teng found out that Min was making money out of the restroom, and thought: If Min could do that, why couldn't he? After all, he owned the use rights of the land before renting it to Lu. Min backed off again when Teng made his claim. Rooted in endogenous Miao values, the original land user's right, which took precedence over the lease contract, was honored. Teng replaced Min charging tourists for restroom use. The village cadres tried again to talk Teng out of it. Teng replied, "Min did not listen to you when he was doing it, then why should I? If your persuasion worked for Min, I wouldn't be doing it now. It was not my intention, but I am not okay with Min making money out of it. If someone is doing it, it should be me. The land is mine!" Unable to dispute Teng's argument, the village cadres could do nothing but leave him alone. I once had to use the tourist restroom. After I was done, I found that Teng's wife was outside the restroom with a lock, waiting to lock it down. She told me that neither Teng nor herself had time that afternoon to sit there collecting fees.

Surrounding the vending stall, tourist restroom, and roadblock issues were individual acts of resistance striving for personal gains of tourism benefits. They used culturally and socially rooted judgements, and resisted in their own innovative ways. These acts were "rightful" resistance from their perspective. Qiu said, "Nowadays the villagers are quite sensible people. They figured that what they did was no big deal, as they did not break any laws to get arrested." He complained that there were no official policies for village cadres like him to rely on, to force the resisters to cooperate whenever necessary. "They [the resisters] are supposed to benefit from what we [the village cadres] are trying to persuade them to do. But whatever we say, they would not listen. It's very upsetting!" He told me that he was a tough man, but he cried twice since he took the position as the village Communist Party Secretary. "While crying, I was thinking: wouldn't it be nice if our policies were like those during the Mao era. If you don't agree, I can take you to the government or the police station to talk some sense into you. If four or five days doesn't work, let's do it for half a month; if half a month is not enough, let's do it for a month." Qiu complained, "Without such policies, it is so difficult for us."

Min was not alone. Roadblocks were a common form of individual resistance in Longcun, as they always triggered an immediate response from the higher authorities. Guo was one of the village rental van drivers. Lu used to allocate some tourists daily to ride in Guo's minivan. But since the merger, such favors were discontinued, and Guo's business suffered. One day, he blocked the village entrance road. As a result, the incoming tour buses changed their tour route to go to other attractions instead of Longcun. Qiu, along with other restaurant owners, reacted quickly. If the tourists couldn't come that day, the food they prepared would be wasted altogether. Qiu talked the other restaurant owners out of confronting Guo directly. "I understand his anger—he worked hard but couldn't get even one single load of tourist passengers," Qiu said. The restaurant owners instead traveled together to the county government office in Tuo River Town to plead for a solution. The county government summoned Guo for a talk. They finally reached an agreement, and the favor to Guo's business resumed.

Another roadblock incident happened when a villager's request for a ride on a tour bus was rejected. Done shopping in Shanjiang on a market day, the villager saw a tour bus on its way to Longcun. The tour bus refused to give him a free ride. He was angry, as such requests were normally accepted and it had become a customary practice. He rented a minivan right away, and followed the tour bus to the village. When the tour bus was ready to leave after the tour, he stood right in front to block its way. He demanded the tour bus operator reimburse the 30 yuan he had paid for the rental van. Qiu was in a meeting and unable to immediately take care of it. He called other restaurant owners, asking for help, "Please go to pull him aside. Don't fight with him. And I will be on my way." Qiu appreciated that they responded accordingly. "At that time, these restaurant owners fairly supported my work," he sighed, and continued, "but not nowadays, and I could no longer turn to them for help."

COLLECTIVE RESISTANCE TO
THE SHOW CANCELLATION

Two months into the merger, without any prior notice, the FFCCTD canceled the show in Longcun. Consequently, tour groups spent much less time, and therefore had much less chance of shopping and dining, in the village. Shocked by the sudden cancellation, the village entrepreneurs reacted promptly. Eight of them, led by Qiu, were on their way to the FFCCTD in

Tuo River Town. Lu heard about their coming, and stopped them outside the FFCCTD. He said, "Don't go in. Let me go to get the person who is in charge of this to talk to you guys. If you go in, it might not do any good." They agreed, and Lu arranged the meeting. Qiu and his fellow villagers were adamant about having the show back. Qiu told me, "That day we made it very clear that they must agree. If not, they'd better give us a good enough reason that we could accept heartily!" As their meeting continued to two o'clock the next morning, they reached an agreement: the FFCCTD would continue to pay the performers' salary, but the village was on its own to find the performers and to organize the show.

Most of Longcun's original performance troupe members were recruited by Lu from elsewhere. They were dismissed by the FFCCTD upon canceling the show, and had already left Longcun. At that time, Qiu's son was performing at another tour site. Qiu called him back, and asked him to bring some of his fellow performers. They managed to find fifteen, who were joined by several more from Longcun. "It was difficult, but we did it little by little," Qiu said. I asked, "Did you ask them at the meeting that day why they all of a sudden canceled the show?" "I did not ask, and they did not explain either," Qiu said. "Whenever we were about to bring it up, Lu cut us off." Qiu continued, "Since they agreed, there was no need for us to find out why. No point of doing that."

Not being offered an explanation, the villagers however had their own answer. They gossiped about how it was related to Yu, or the "old fox" as they called him. As the former vice head of the autonomous prefecture, Yu was a local celebrity. He often featured on national and local television stations and newspapers. To the outsiders, Yu was a respected scholar and educator who dedicated his life to salvaging Miao culture; to many of the local people, he was an ambitious and cunning businessman, one with special political leverage. Yu had strong connections to those on the top of the local political hierarchy, many of whom were once promoted by or received favor from him before his retirement. Yu was the owner of the Shanjiang Miao Museum. It was also merged into the FFCCTD, and he was on its board. As mentioned earlier, his museum and Longcun were the two attractions featured by the same tour route, between which the one-day tour time was split—whoever kept the tourists longer got more out of their wallets. Upon the merger, Yu added a show performed at his museum site. Tourists had little desire to watch two similar shows on the same day: one must be cut, and then it shall be the one in Longcun.

While Longcun was successful in reinstituting its show, the success was short-lived. It was discontinued soon due to the lack of appropriate management. Jia's grocery store was by the main village road close to the parking lot. Jia was from Tongren in Guizhou. She moved here after marrying into Longcun. The largest in the village, the store seemed busy during summer peak season. Frolicking village children ran in with small change in their hands for candies; thirsty drivers and tourists stopped by for chilled water and drinks. It was also a gathering place: bored drivers hung out while their tour groups were touring the village; and strayed tourists waited for reunion with their tour groups. However, Jia didn't think that her business was nearly as good as those at the Shanjiang Miao Museum. One day she went to Shanjiang to replenish popsicles. She asked the wholesale store to give her one carton. Another buyer heard and said to her, "Only one carton? For me, it's always a dozen." It turned out that his store was at the museum site. Noticing that the popsicles Jia was buying were those priced at one yuan, he teased her, telling her that she should get some more expensive ones as he did.

In contrast to Longcun, the performance area in Shanjiang Miao Museum was crowded with tourists. The museum was housed in a compound, and the performance area was in its courtyard. It was the former residence of Yunfei Long, a well-known Miao rebellion leader and later a high-ranking officer of the Guomindang in the 1930s. When the rebellion was put down in the 1950s, Yunfei Long was killed, and members of his family who were involved in the rebellion were persecuted. The township government confiscated the house. While Yu is a native of Ala Town, his wife is from Shanjiang Town and one of her relatives was the township government official. Upon his retirement in 2001, Yu moved to Shanjiang. In 2002, Yu leased this house from the township government for 50 years, and opened the Shangjiang Miao Museum. When I first visited the museum in 2005, there were only exhibits, no performance show.

Yu actively sought to expand his museum business, which encountered resistance from the local residents. The land that Yu used as the museum parking lot belonged collectively to Shicun village. When Yu started to charge a parking fee for tour vehicles (40 yuan per bus, and 20 yuan per sedan), the villagers were fiercely against it. Without leaving the opportunity to Yu, they went to guard the parking lot and charged vehicles for parking. Yu also had his eyes on the land between the parking lot and the museum entrance, where there used to be a state-owned grain shop. With the grain shop closed down, the land had been returned to the village as a collective property for many years. Yu offered the villagers 600,000 yuan. The villagers counteroffered with a prohibitive high price of 16,000,000 yuan. When I revisited the museum in 2013, I saw the villagers kept their vending stalls lined up between the parking lot and the museum entrance, continuing doing business with the tourists.

Yu's ambition exceeded his museum. In mid-July of 2010, he was joined by the developers of seventeen countryside tour sites, and they set up the

Golden Phoenix Tourism Company (GPTC) in Tuo River Town. Yu was the chairman of the board. They initiated a "three-day pass": by spending 148 yuan to purchase a "three-day pass," tourists could visit as many of the eighteen sites as possible within three days. The overall profit was distributed to the individual site according to the number of tourists it received. Tourists could also choose to buy an individual ticket for each site, with the price ranging from 48 to 68 yuan. That summer, the GPTC's salesmen were in the streets of the *Gu Cheng*, handing out promotional flyers to tourists. The company existed for two months before it fell apart. Some, including Lu, suspected that Yu was behind the merger and the establishment of the FFCCTD, and the GPTC was his first attempt.

INDIGENOUS FORMS OF EVERYDAY RESISTANCE

Facing external pressure from Lu and later from the FFCCTD, everyday forms of resistance in Longcun involved trespassing, squatting, roadblocks, sabotage, gossip, and quarrels. Similar to the everyday resistance in Sedaka discussed by Scott (1986a), peasant resistance acts in Longcun required little coordination, and were continuous daily tactics. These tactics were representative of everyday peasant resistance in Fenghuang. They were indigenous responses arising from similar impacts of tourism development, and were unlikely to disappear altogether so long as the local tourism system remained exploitative and inequitable

Scholars (Colburn 1989a; Haynes and Prakash 1992; Scott 1986b) argued that everyday resistance rarely helped the subordinate to escape from domination. As Scott (1986a: 30) summarized, the possible consequences of peasant resistance might marginally alleviate exploitation, might amount to a renegotiation of the limits of appropriation, or might change the course of subsequent development, but rarely helped bring the system down. In Longcun, the incidents triggered by the allocation of vending stalls posed no fundamental threat to the existing hierarchy. Neither did the collective complaint regarding the sudden cancellation of the show aim to challenge the established tourism system.

These resistance acts opposed neither tourism itself, nor the way that tourism was developed. Rather, they were intended to create opportunities for the resisters to negotiate with the developers and local authorities for better access to tourists, the source of tourism benefits. They presented a constant process of testing, negotiation, and renegotiation of relations between the subordinates (the villagers) and the superordinates (including the county government and private developers such as Lu). For both sides, there was a never-ending attempt to probe the limits, to exploit any possible cracks, to see precisely what could be tolerated, and to seize each small advantage.

Tourists as Scapegoat

Typical forms of resistance feature little coordination or planning, self-help, and avoidance of direct affront to authority (Scott 1986b: 1). Scott (1989: 25) considers "the pervasive use of disguise" to avoid direct affront to authority as "the most striking characteristic of normal resistance by subordinate groups. He indicates that such disguise could be either the concealment of anonymity of the resister if the message is clear, or the ambiguity of the message if the messenger's identity is clear. However, in Longcun we see yet another type of disguise at work.

Generally, the resisters in Longcun employed two types of disguise: 1) a clear message delivered by disguised messenger(s) to the dominant (the private developers and political authorities); and 2) a clear message delivered by identifiable messenger(s) to a scapegoat (tourists) of the dominant. The second type was more common in practice than the first. The first type resembles what Scott characterized; the second differs in that the real target is concealed, while both the message and messenger(s) are clear. When the resistance was directly aimed at the dominant, it was covert (e.g., sabotaging Lu's drums, damaging the water pipe). When it was overt, it tended to involve face-to-face confrontation with the tourists (e.g., blocking their entrance, and charging them for restroom use).

The latter demonstrates a certain tactical wisdom, as tourists are the relatively less powerful agents in the local touristic system. Direct physical confrontation with the developers and/or political authorities involved more risk, and might not have worked as effectively as with the tourists who were the source of economic gains for the dominant. As a form of "avoidance protest" (Adas 1981) in taking out their frustration on the tourists to express their discontent, Longcun villagers minimized the costs associated with direct clashes with, and at the same time guaranteed immediate attention from, the developers and local government. To some, it might still be interpreted as direct resistance to the dominant; however, this was not direct in the sense of initiating face-to-face physical confrontation with them.

Policy Gaps as Weapon

Li and O'Brien (1996) discuss policy-based "rightful resistance" in rural China, in which case relevant policies are the weapon for peasants to resist lawfully. In Longcun, on the contrary, the peasants utilized gaps in existing policy to persist in their resistance. In this case, the absence of relevant state policy worked to their advantage. They tacitly utilized the fact that there was no state policy for the enforcement of tourism development: they did not take orders; neither did they break any laws. The village cadres may legally resort to heavy fines or physical force in implementing the family planning policy, but not in pushing forward their agenda in tourism planning. Due to the absence of a policy basis for enforcement, they had to rely solely on persuasion. Such persuasion alone unlikely worked to make the noncompliant compromise. The village cadres complained of the difficulty of their job, and attributed it to the peasants' backward mind (*sixiang luohou*), and their lacking of culture (*meiyou wenhua*).

Unlike township and county officials, village cadres are not "official" because they are not members of the state bureaucracy. Village cadres' position in local society as both fellow villagers and kinsmen made them more considerate of the interests of other villagers. They were often caught between what Colburn (1989b: xii) called "the upper and lower jaws." Squeezed from above and below, they shifted their role between resister and middleman between their fellow villagers and the dominant. Their actions varied, depending on how great their own personal interests were at stake. While facing roadblock incidents, Qiu was more in line with the dominant, despite his sympathy for his fellow villagers who conducted roadblocks which could devastate his restaurant business; when the village encountered the sudden cancellation of its performance show, he switched sides as shared interests prompted him to lead his fellow villagers to confront the dominant.

DOMINATION AND RESISTANCE

The tourism development model adopted by the county government—one that relied heavily on the privatization of natural and cultural resources—has been deepening social inequality in Fenghuang. It contributed to the rise of everyday resistance in local rural communities such as Longcun. "Where there is power, there is resistance," says Foucault (1984: 95). The interaction of domination and resistance constitute much of the lived experience of economic injustice, as shown in the case of Longcun. The local government and private developers, as the alliance of money and power, jointly appropriated profitable resources and maximized their personal gains through promoting tourism development. Their domination encountered day-to-day peasant resistance in the forms of trespassing, squatting, roadblocks, sabotage, gossip, and quarrels.

It is important to specify exactly what was being resisted. Longcun villagers' resistance was not in opposition to tourism: they generally welcomed the opportunity of developing tourism. Their everyday resistance, prosaic but constant, was carried out over the years during an elite-directed and capital-intensive tourism boom, fighting to be included in this otherwise exclusive development process. In responding to various situations, they made rational choices among possible trajectories of action. If the resistance was direct, it was not open; if the resistance was open, it was not direct. As the dominated, they worked the existing system to their advantage: they cleverly utilized policy gaps to persist in their resistance; and they strategically used tourists as the scapegoat to resist the dominant. No matter how marginal and temporary the results of their resistance might seem, they were successful in affecting the various forms of exploitation that they confronted. What they accomplished was by no means trivial, as it almost always resulted in gaining opportunities for negotiation. It was significant in the sense that it narrowed the options available for the dominant regarding tourism development.

Everyday resistance in touristic Longcun is part of the ubiquitous struggle against the effects of the capitalist development in Fenghuang. Similar cases include Shicun village's resistance to Yu and his effort to expand the museum business. They reflect widespread struggles elsewhere in contemporary China. Today, many peasants are defending their interests in increasingly neoliberal China. It is important to understand from a ground level, a scale to which analysts have paid limited attention, if and how their everyday forms of resistance alleviate exploitation, amount to a renegotiation of the limits of appropriation, or, in rare occasions, even change the course of subsequent development directed by the elites.

Agreeing with Scott (2005: 398), I believe that "no abstract force, collectivity, or system ever arrives at the door of human experience, except as it is mediated by concrete, particular human carriers." To portray the resisters and unravel how particular resistance incidents have configured among them is indispensable to a nuanced understanding of resistance and of the transformation of those societies under capitalist penetration, as in this case through tourism development. Analysis of tourism has much to contribute to the understanding of everyday resistance; and a more detailed treatment of everyday resistance is vital to a fuller understanding of tourism-power relationships.

NOTES

1. Zweig (1989) elaborated on peasant resistance that resulted in the post-Mao changes especially regarding land issues after collectivization during 1960s–1980s. More recently, other scholars have examined collective protest and violence and policy-based resistance as new forms of peasant resistance in China. For example, political scientists Li and O'Brien write extensively about policy-based "rightful

resistance" (e.g., Li and O'Brien 1996; O'Brien and Li 2006), and based primarily on archives, historian Walker (2006, 2008) also focuses on the peasant collective protest and policy-based resistance in China's post-socialist period.

2. Resistance studies (represented by Scott's work) have been criticized, among other issues, for "romanticizing" (Abu-Lughod 1990) and "fetishizing" (Kellner 1995) resistance, for "essentializing" subordinates (O'Hanlon 1988), for the missing link with collective action (Escobar 1992), for the ignorance of socio-economic differentiation among peasants (White 1986), for ethnographic thinness (Ortner 1995), for "inherent explanatory limitations of the resistance concept" (Brown 1996), for "translating the apparently trivial into the fatefully political" (Sahlins 1993), for the lack of utilizing a psychological approach to address the problem of an actor's motivation (Seymour 2006), or for its overemphasis on the actor's experience of personal domination while bypassing the experience of impersonal domination shaped by values, traditions, or the symbolic universe (Ho 2011).

3. Through examining their embroidered textile (re)production, cardamom cultivation, and involvement in trekking tourism, Turner and Michaud (2009) and Turner (2012a) demonstrated how the Hmong in Vietnam resisted unwanted levels of market dependency through their selective involvement of trades. Turner (2012b) discussed how the Hmong farmers in Vietnam have adapted to, circumnavigated, or resisted state-sponsored agrarian change, and revealed their culturally based adoption and resilience to both the market and social integration practices of the Vietnam state. Similarly, as an ethnographically rooted take on infrapolitics and everyday resistance, Michaud (2012) looked at agricultural practices, clothes production, technology adoption, and education among the Hmong in Vietnam, and suggested that they were being tactically selective about modernity. In one of his earlier works, Michaud (1997) explained from a perspective of cultural resistance why tourism had so little impact on a Hmong village in Thailand, even after several years of daily contact.

4. Partial material in the remaining chapter was previously published as "Protesting Power: Everyday Resistance in a Touristic Chinese Miao Village," *Journal of Tourism and Cultural Change* 13(3): 225–243. I thank the publisher Taylor & Francis Group (www.tandfonline.com) for the permission to use these materials.

Chapter Seven

Competition and (In)equality

The Rise and Fall of Village Family Restaurants

After visiting Ting's (Jing's second younger brother) family in Shanjiang, I returned to Longcun.[1] I ran into Zhi, the former village Communist Party Secretary. It had been a year since we last saw each other. Zhi invited me to have lunch at his house. On our way there, he told me that he recently started a restaurant business at his own house, serving food on order for the *san ke*, rather than standard meals for tour groups. I asked, "What happened to the *miaojia* restaurant?" Zhi said, "It was closed. Not much business." The *miaojia* restaurant was one of the six family restaurants in Longcun, jointly ran by Zhi and four other villagers. It was close to the performance area, using the land of one of the four villagers. I dined there once two years ago in 2011, and still remembered the busy scene back then.

Zhi's house was not on the village's tour route. Zhi made several wood signs and hung them on the trees along the way between the performance area and his house. On the sign was written "*laozhishu fanzhuang huanying nin*" (The Former Communist Party Secretary Restaurant Welcomes You), with an arrow indicating the direction (Photo 7.1). Soon after we got there, Zhi had the first group of customers for that day. A family of three generations; they drove to Fenghuang for vacation. They ordered ten dishes. While Zhi was busy killing, plucking, and cutting a rooster, I helped prepare other food materials and chatted with the guests. Chen, in his early sixties, was the head of the family. He owned a family-run factory in Xuzhou, Jiangshu Province. Many workers in his factory were migrant laborers from Fenghuang. "I could not bear living here, not even for one day," Chen confessed, saying, "The conditions are too bad." Chen's daughter asked me if there was another bathroom after she found the outhouse in the pigpen.

While the guests were enjoying their meal, Zhi put on his eyeglasses and took out an abacus, calculating the total charge to put on the bill. It was 251

Photo 7.1. By the trail between the performance ground and Zhi's house, there was a wood sign hung on a tree, on which was written *"laozhishu fanzhuang huanying nin"* (The Former Communist Party Secretary Restaurant Welcomes You), with an arrow indicating the direction. (Photo by Xianghong Feng, 2013).

yuan. On the best day, Zhi had gross earnings close to 1,000 yuan, but those days were rare. Nevertheless, he preferred food on order for the *san ke* to the standard meals for tour groups, saying the latter was "too competitive." Without exception, the six family restaurants in Longcun had all collaborated with tour agencies, and specialized in providing standard meals for the packaged tour groups. During the past two years, there had been increasingly fierce competition among them. This chapter focuses on the rise and fall of these family restaurants, and explores how the traditional peasant worldview affected peasant entrepreneurship in market economy.

"LIMITED GOOD" AND "COMPETING TO REMAIN EQUAL"

The gods of wealth in China are believed to be capable of transporting rather than producing goods, making one family prosperous by ruining the other (Li 1948: 11). The ideology of the "gods of wealth" in a sense reflects the view of "limited good" (Foster 1965) that sees social division of wealth holistically: the accumulation of wealth for some means wealth being taken away from others at the same time (Martin 2015: 77). As an integral aspect of the peasant moral economy, the image of "limited good" is viewed as still being central in many contemporary peasant societies (Van der Ploeg 2008, cf. Tucker 2010: 935). According to Foster (1965, 1972), peasant communities approximate a closed system, and are ruled by the Image of Limited Good. In a traditional peasant village, no individual can get ahead at the expense of fellow villagers. However, one can get ahead by obtaining wealth existing outside the village system (e.g., working as a migrant wage laborer). Though envied, it is not being seen as a direct threat to community stability, as it costs no one in the village anything (Foster 1965: 306).

The competition for the "limited good" is "competing to remain equal." Peasant societies are characterized by a constant competing to remain equal, as initially described by Bailey (1969, 1971), Foster (1965) and Scott (1976). Competition normally occurs among equals or those almost equal, as a gap in social power and status distances people and disables competition.

Bailey (1971: 19, 23) states, "People remain equal because each one believes that every other one is trying to better him, and in his efforts to protect himself, he makes sure that no one ever gets beyond the level of approved mediocrity. . . . Skills and energies go into keeping people in the place that they have always been: they run hard in order to stand still. It is the kind of world that swamps heavily upon change and innovation." Peasant sociality is thus resistant to entrepreneurial innovation and development. According to Tucker's (2010) research on tourism in Goreme of Greece, "jealousy" was frequently mentioned by the villager entrepreneurs as a prominent issue attributing to increased internal conflicts, who complained that some were willing to take a loss themselves to prevent others from getting ahead. Hence, "jealousy" in such context works as a mechanism for equality. It is a self-defensive reaction, to restore the original equality and fairness rather than to advance one's own position in society.

There is an academic version of this general way of viewing the world as "limited good," including a school of thought in political economy known as dependency theory (Frank 1966), and the "zero-sum" notion in Wallerstein's (1974) "world-system" analysis. Nash (2007) contrasts the "unlimited good" (the open-system model long promoted by economists) and the "limited good" or "zero-sum" (the closed-system model), and points out that replacing the "unlimited good" with the "limited good" worldview helps us explore the alternatives beyond neoliberalism market economy for equitable and sustainable development. Nash (2007: 35) argues that the empirically derived model of the "limited good" that Foster constructed to explain peasant rejection of modernization also cultivates conservation practices that benefit sustained development. As she (Nash 2007: 35–36) states, "This more balanced outlook in which development for the collective good with redistribution of the rewards of production promotes sustainable practices provides an alternative to the specter of impoverishment fostered by unlimited growth benefiting only a few." Trawick and Hornborg (2015) further discuss the differences between the "unlimited good" and "limited good" worldviews, and argue: in order to obtain equitable and sustainable development, a shift toward the closed-system of "limited good" worldview is necessary.

RISE AND FALL OF THE FAMILY RESTAURANTS

Just as Nash (2007: 36) said, "I hope to rescue Foster's insights on the peasant worldview and what that may offer for those concerned with a sustainable future for world population." In the following, I revisit the viewpoint of "limited good" and "competing to remain equal" in a tourism context, and discuss the cooperation and competition among the family restaurants in Longcun. It is to better understand the cultural elements behind peasant entrepreneurship in contemporary China, and shed light on worldwide issues of collective prosperity and equitable development.

Cooperation

Qiu opened the first family restaurant, *nongfu* restaurant (Farmer's Restaurant). It was in 2009. Three months later came the second one. Half a year later, it turned into three. By 2011, there were six of them, including *nongfu*, *miaojia*, *yiqing*, *shanli renjia*, *huiyou*, and *xiaoyingbin* restaurants. *Yiqing* restaurant (Friendship Restaurant) started its business during the Labor Day in 2010, jointly run by four households, including Jing, Jing's cousin, and Mi. When I became interested in Fenghuang's countryside tours, I participated in a tour group and experienced a one-day Miao village tour, my first visit to Longcun. The tour guide took us to lunch at *yiqing* restaurant. The dining area was a roofed wood structure placed with eight round tables. It was right outside one of the co-owners' houses. The kitchen was in the house, equipped with four stand-alone coal stoves in addition to its original built-in firewood stove, a small scale of mass production of meals. The restaurant had been three months into business. The dining tables were packed that day, and Jing, Jing's cousin, and Mi were busy in the kitchen. However, a few months later, Jing had a fight with his cousin over the management of the restaurant. Jing withdrew, and opened *shanli renjia* restaurant in his own house. He and his cousin did not speak a word to each other for two years. *Yiqing* restaurant made a profit of approximately 100,000 yuan by the end of that year. Jing received 20,000 yuan, and each of the other three households obtained 26,000 yuan.

The following is the excerpt from my interview with Mi in the summer of 2011.

Feng: After Jing withdrew, is the profit higher this year?

Mi: If the tourists are as many as last year, we three households should make bigger profit than last year. But so far, it is not as good as last year.

Feng: Why have you three households run it together? More people then less profit, isn't it?

Mi: It won't work if not enough people. The customers are tour groups, and they are a lot. The reservations usually come in around 8–9 am, requesting lunch to be served around 11:30 am. With nine dishes for each table, it won't work if not enough people.

Feng: So, it is mainly a problem of manpower, not funds?

Mi: See, we must at least have three households. But if there were only one person from each household [to work at the restaurant], we would have to hire people. Otherwise, it won't work.

Feng: Then why doesn't one household run it on its own while hiring as many hands as needed? Won't it be higher profit this way?

Mi: Whether we hire depends on the situation. When there are many tourists, it makes money; when less tourists, it does not pay off. For example, in off-season, our own people don't care about wages, and we together spend our time here [at the restaurant]. If we hire people when only a dozen guests, no profit at all, let alone the cost for the food materials.

Feng: How about relying on your own when there are fewer guests, and hiring people when more guests? Not necessary to have multi-households for joint management, right?

Mi: For our three households, if two people from each household [work at the restaurant], we have six people. Even if there are a couple hundred or three hundred guests, we don't need to hire people.

Feng: How does the partnership work among the three households? Did each household invest the same amount of money?

Mi: Yes, the same amount of money.

Feng: How about the land [used for the restaurant]?

Mi: It belongs to one of the other two households, and we pay no rent. The water is mountain spring, so no water bills. If it is in town, there are taxes. In our village here, neither taxes nor rent.

Feng: Does the household who volunteered the land receive more when the profits are divided among the three households?

Mi: That is among us, and we have agreement. If he does not accept it, he won't accept it even if we give it to him. He is willing to use his land for our restaurant. No string attached.

Jing was the head of *shanli renjia* restaurant (literally means a family restaurant among the mountains) (Photo 7.2). Despite being almost illiterate, Jing, an outgoing man in his mid-forties, did well with his business. He was in charge of everything. His older sister prepared food and washed dishes, his brother-in-law cooked, and his wife helped after completing the household chores. During school break, his daughter returned from the middle school in Tuo River Town and also helped at the restaurant. Jing had no free moments. He supervised the helpers, replenished produce and meat from the market, swept the floor and cleaned the bathroom, visited with tourist customers and showed hospitality, and networked with tour guides and anyone who might bring potential customers. When there were too many guests, Jing hired one or two villagers as helpers with a daily wage.

Photo 7.2.   In early morning before the tour groups' arrivals, Jing sat by the entrance of his family restaurant, to receive phone calls from tour guides and write down the reservations they made for that day. (Photo by Xianghong Feng, 2014).

When San Ge decided to resume his life as a migrant worker in Hangzhou after his son's accident, Jing asked him to stay and join his restaurant business. San Ge turned it down. He was concerned about working for Jing. He said to me, "I am worried that we may not get along. He is my older brother, and I won't have much freedom if I work under his nose every day." Jing did not give up, and kept trying to talk him into it over phone calls while he was in Hangzhou. With some savings, Jing wanted to purchase a van for the restaurant food-supply-replenish trips, and San Ge was the only one in the family who had a driver's license. In early August 2012, San Ge took a few days' leave from his factory job in Hangzhou and came back to visit. Jing convinced him that he needed his help. When San Ge was leaving for Hangzhou, he planned to quit his job as soon as he received his wage for that month. He needed to be back in time to help Jing shop for a van before the National Day (the first day of October), when they expected to receive a large number of tourists due to the weeklong holiday. Right before National Day, Jing bought a van for 70,000 yuan. It was an eight-seat stick-shift MiniPassenger Van, with a built-in GPS system. It was a slight upgrade of the minivan San Ge had before, thus less likely to catch the traffic police's attention to check for a commercial license.

In the morning, San Ge helped in the restaurant kitchen; during the day, he did rental van business to transport the *san ke* between the village and Shanjiang; in the evening, he, with Jing or alone, drove to the agricultural wholesale market in Tuo River Town to replenish food supplies for the restaurant. At the end of each day, San Ge turned in the earnings from the rental van business to Jing, which was put together with the earnings of the restaurant. The overall monthly profit was divided into three shares for Jing, San Ge, and their older sister. For the latter two, 2,000 yuan was deducted from each of their shares to pay Jing, which was considered as their contribution for the van purchase. During non-peak season, San Ge normally received 2,000–3,000 yuan per month. It was more during peak season, such as in July and August. One day in August 2013, I accompanied San Ge to drive tourists from the village to Shanjiang. San Ge swung by the bank and deposited 3,000 yuan cash. He went to the bank twice to make deposits that month. And that month the profit of the restaurant was close to 60,000 yuan.

San Ge's house was a stone's throw away from Jing's. Jing offered money to renovate San Ge's house, to use it as an expansion of his restaurant. San Ge declined. San Ge was saving up money, and finally did the renovation and opened his own restaurant, *juxiang lou*, in early 2014. When I was helping at *juxiang lou* that summer, I asked San Ge, "Did Jing agree that you would open your own restaurant?" San Ge said, "He told me not to. He wanted me to just make rice liquor to sell." "Was it because he was worried that you might lose money?" I asked again. "No. He was worried losing me as helper at his restaurant," San Ge replied. He continued, "There is no competition between us. My restaurant serves food-on-order for the *san ke*, and his provides standard meals for tour groups." San Ge stopped the rental van business, but continued driving the van for Jing, and he was responsible for replenishing food supplies for both restaurants. Since he learned how to make rice liquor, he provided rice liquor to Jing's restaurant. "I sold it to others at 120 yuan per barrel, but I only charge my brother 60 yuan, with a profit of 20 yuan," he told me.

According to Nash's research (2007) in Chiapas in Mexico, those who tested out new modes of making an income usually sought associates in a co-operative that shared the risk of making a profit. Similarly, the family restaurants in Longcun, such as *yiqing* and *shanli renjia*, were usually jointly run by several households. This type of co-management was built upon mutual aid in daily life between households related by kinship. Trust-based relationships in networks of kin are used to reduce risk. Labor is the poor's greatest asset (Moser 1996, cf. Gonzalez de la Rocha 2001: 81). Compared to hired labor, they were more willing to rely on the "free" labor of their partners. The use of kinship ties in the joint ownership restaurant business benefited from sharing both resources (e.g., the venue) and costs, which reduced risk and lowered barriers for a start-up, despite that reducing risk likely meant reducing average earnings per co-owner. This reflects one important feature of the peasant mode of production: risk reduction, with safety (not profit) as the priority.

For the vast majority of human history, the economic system was run on noneconomic motives (Polanyi 1944: 46). Peasant economy obeys its own principles, which could not be simply interpreted by economic models which are mainly based on the indicators such as price, supply and demand, and profit, as Chayanov (1996) argues. In her discussion of rotating credit societies in traditional China, Martin (2015: 49) cites Fei (1939: 268), and states that these credit societies were only allowed to serve those needs that did not interfere with community solidarity, such as marriage and funeral ceremonies. They were not for productive purposes such as starting a business or buying a piece of land, as using the accumulated profits to increase stratification within the community was not permitted.

Peasants worked hard year-round, driven by self-sufficiency for the entire household, not by pursuing maximum profits in the market. They tended to depend on their own "free" labor rather than hired labor. Networks of both shared costs and labor reciprocity are argued to be integral to peasant communities in that they constitute "the moral economy of the peasant" (Scott 1976). In peasant studies, there had been tit-for-tat argument on whether the peasant economy was dominated by "morality" or "rationality" (Scott 1976 vs. Popkin 1979). Through the case study of a *gong fen* (work points) system implemented by the Langde Miao village in Guizhou of China during its tourism development, Li (2008) argues that the *gong fen* system reveals not only survival principles (i.e., mutual benefits, fairness, and safety first) as emphasized by "moral economy" (Scott 1976), but also power conflicts and competition for personal gains as outlined by the "rational" peasants (Popkin 1979).

Competition

While cooperation and solidarity are essential to clan-based communities, there is increasing competition and friction among those involved in the tourism business. Through the case study of Pingan Village in Longsheng County of Guangxi Province in China, Huang (2006) explores the conflicts among multiple stakeholders in an ethnic tourism context. Pingan was one of the villages with the fastest tourism growth and highest tourism profits in the ethnic minority area in Guangxi. Meanwhile it had the most conflicts, including those among the villagers. To compete for tourists, the villagers often resorted to price wars. Disorderly competition led to hostility.

Similar situations were found in Longcun. It was no secret that the restaurants in Longcun competed fiercely with one another. When I visited Ting's family in Shanjiang, his wife commented on Longcun's internal friction, particularly referring to the tension among these restaurants. Tour guides normally made a reservation with one restaurant ahead of time to arrange the dining for his/her tour group. But sometimes, incidents like this happened: while a tour group was passing a restaurant on the way to a competing restaurant where meals had been prearranged, this restaurant dragged the tourists in and tried to persuade them to eat there instead. The tour guide couldn't do much with the situation if the tourists chose to stay. Thus, the restaurant with the reservation not only lost customers, but also wasted food that they had already prepared to serve.

For the villagers whose restaurant business did less well, their state of mind was not improved by the sight of flocks of tourists passing by their houses with all the empty dining tables and going to their neighbors'. The daily flow of customers became a public display of individual restaurant's capability. In a sense, face (*mian zi*) mattered more than the actual total profit. Each restaurant tried to keep track of all counts of tourists: tourists who dined at its own restaurant, tourists who went to other restaurants, and the total tourists in Longcun. In the midst of outcompeting each other for more customers, they somewhat lost their original focus on maximizing profit. Soon, price wars broke out. Their frustration and anger heated up along with it: due to declining prices, they had to work much harder to prepare more food and serve more tourists for the same (if not less) profit. Due to the high kickbacks demanded by the tour guides, the profit margin for the restaurant shrank. The restaurant charged 10 yuan per person per meal. The tour guides collected a 50-yuan meal fee from each tourist in their tour groups, gave 10 yuan to the restaurant, and kept 40 yuan to themselves. The costs for the restaurants included food supplies and the transportation to replenish them, cooking fuel, labor to prepare meals, attend guests, and clean tables and the floor, plus the free meals offered to the tour guides and tour shuttle drivers.

In early 2012, Qiu called for a meeting among the restaurant owners. He proposed to set the meal price at 10 yuan until the end of 2012. All agreed. Three restaurants implemented it, and noticed much less business. It turned out that the others charged a lower price of 8 yuan. Hua, who regularly took tour groups from Jun's travel agency to Longcun, commented, "What happened is that they asked for the same price, but somehow ended up with fewer customers. Then, they figured that if they drop the price, they would have as many customers." She continued, "If other restaurants confront them, they would go: you already have so many customers, and why wouldn't you let me lower my price? If I did so, I might have as many." "They weren't thinking straight," Qiu said. "What they had in mind was: if you had more customers, I must keep up; to keep up, I must lower the price."

Based on the market economy principles, healthy competition should focus on improving the product and service quality and enhancing its difference. This requires a certain amount of capital support. In addition, the costs are higher, and the earnings cycle is longer than touting and other short-term business activities. Thus, similar to the competition among the residents in Langde Miao village (Chen and Kuang 2009), the family restaurant owners in Longcun took actions for short-term benefits. They were all engaged in touting: if others adopted this strategy, whoever was not doing it would suffer a loss in earnings. Fighting for tourists caused a gradual rift and increasing conflicts among the villagers.

Double "Backwardness"

"More developed, more conflicts!" as Mi commented. Mi claimed that in the old days, their conflicts only occasionally erupted among the villagers over the boundaries of woods and cropland, or between the village cadres and other villagers over the implementation the family planning policy. Increasing conflicts in Longcun came along with its tourism development. These conflicts initially occurred between the villagers and the outsiders such as Lu, tour bus drivers, and tour guides, as discussed in previous chapters. With more villagers involved in the tourism business, it gave rise to internal conflicts due to competition, like those among the family restaurants. Mi blamed the increasing conflicts on the "lack of culture" (*meiyou wenhua*) of his fellow villagers, and said, "Countryside people are short-sighted, and egalitarianism is serious in our village. That's right, egalitarianism, that's the problem!" Mi pointed out the Miao egalitarian morality, and considered it a drag on the village development. The village cadres and Lu often complained to me that it was very difficult for them to push forward tourism development in Longcun, as the villagers were too "backward," too ignorant, and too unreasonable.

I recalled that during my trip to Wen's "Wanrong River Rafting" site, I chatted with a group of raft-operators. One of them asked me whether Miao was an ethnic group with the most solidarity in the world. When I returned from the rafting site to Lacun, I visited Wen. During our conversation, Wen talked at length about dealing with people, the biggest challenge for him to manage this business. Wen said, "As the Miao, we are a unity; but as individuals, we are stubborn, with our own thinking and behaviors." He continued, "If other groups bully us, we unite. We don't need know each other personally. We can tell if someone is one of us as soon as he/she speaks. We might be strangers, but as long as we speak the same language, we fight for one another. This is our habit. But when it comes to doing business, as individuals we Miao only consider one's own benefit. The Miao are conservative, not open. . . . When not open, it is very hard to do business. You see, it is very difficult for a Miao like me to run a business with my fellow Miao."

I often heard similar complaints during my time in Fenghuang. Ru, as a Miao official, often talked about the "backwardness" of the local Miao peasants. She once used an anecdote as example. Liangdeng is another Miao village in Shanjiang. With no motor road, it took over three hours to walk to its market town, more remote than Longcun. The county government sent a Poverty Task Force to build a road, and asked the villagers to unload the sands. The villagers however demanded compensation for their labor. Ru said, "These villagers were incredibly unreasonable. The government offered help, but they demanded money. They did not welcome the government to build the road for them, saying that if there was a road, it would be easy for the family planning people to come, and their pigs might be taken away by outsiders."

These perceptions to a certain degree reflect the mainstream belief of attributing the root of poverty to the "lack of culture." Why did the seemingly well-intentioned initiatives, which aimed to improve the villagers' living standard as seen from the perspective of the elites, encounter resistances from the villagers? If viewed in the light of the "Image of Limited Good" (Foster 1965), the conservativeness was not due to backwardness, but because of individual progress being seen as the threat to community stability. Thus, the villagers would try every means to prevent change in the status quo. Such competition derived from the "limited good" viewpoint is the "competing to remain equal."

As Cohen (1993: 166) argues, the creation of a negative depiction for China's peasants was related less to the circumstances, potential, and culture of its rural inhabitants, and more to an elite antitraditonalism that formed a moral claim to political privilege and power. Being rural and ethnic, the Miao peasants bear the double burden of "backwardness," an everyday reality they live and breath. I once took Liang, Hong, and their toddler boy out for dinner in the *Gu Cheng*, and I got a taxi to take us back afterward. After dropping them, the taxi driver asked me how I got to know them and whether they were from Shanjiang or Laershan (both with high density of Miao). "No, they were not," I replied, then asked, "but how do you know they are Miao?" He said, "In Fenghuang, the town people and countryside people have different accents; even if both from countryside, the Miao and Han are still different." He was puzzled how I, an outside city woman, was friends with the local Miao peasants. Qiang, a Han peasant, was one of my informants and friends, who then had a temporary job at the long-distance shuttle station in Tuo River Town. Whenever I misunderstood him, he always teased and said, "Why were you like those Miao, who just couldn't get what people say?!"

Yiqing restaurant was closed in 2013. Mi complained that it was difficult to cater to tour groups: whoever lowered the costs the most to give tour guides the highest kickback had the most customers. After *yiqing* was closed, Mi looked after his two-and-a-half-year-old grandson, so that his son could focus on doing a tourist photograph business near the performance area in the village. As mentioned in the opening of this chapter, *Miaojia* restaurant closed about the same time after staying open for two years, due to the lack of customers. Then Zhi started *laozhishu* restaurant at his own house, specializing in food on order, catering to the *san ke* rather than tour groups. Zhi said to me, "I don't want to do group meals anymore. The meal price is too low, and not much money to make." Due to the distance of Zhi's house to the tour route, Zhi did not have much business. One day San Ge was giving me a ride to Tuo River Town. While driving through Shanjiang, he saw people working on building a house right by the road. He mentioned that Zhi had to come to town for a construction job like this to earn wages because of the hardship of his own restaurant business. By the summer of 2014, there were only three restaurants doing group meals including *shanli renjia*, *huiyou*, and *xiaoyingbin*.

TOURISTS AS THE "COMMONS"

In Longcun, the number of tourists of a certain day was fixed, and there were budget limits for most of them. Thus, it might be considered as the "limited good," since the benefits from the tourist business could be perceived as wealth drawn from the village (a closed-system), different from that obtained as wage-laborer from the outside world (an open-system). So, tourists could be perceived as "common-pool" resources in a tourism context. Daily conflicts stirred by envy were aggravated by the growing wealth distinctions in a community where there had been shared poverty during pre-migration and pre-tourism time. While discussing "the tragedy of the commons," Hardin (1968: 1243–1248) states, "Each man is locked into a system that compels him to increase his herd without limit—in a world that is limited." Hardin's statement in a degree echoes Foster's concept of the Image of Limited Good. These restaurants could do nothing to increase the total number of tourists in Longcun, their potential customers, in a given day. What the tourists spent on their meals were the "limited good" to them. Therefore, the meal consumption of the tourists, as the "commons," might become "overgazed" by these restaurants through lowering their meal quality due to price wars.

Longcun's case elucidates Martin's finding (2015: 60) that in late traditional into modern China, social considerations took precedence over the strictly economic ones. In an acquaintance society such as Longcun, the villagers had frequent daily interactions and "face" (Hu 1944) matters. For the villagers who ran family restaurants, winning more customers might be more of a matter for "face" than for profit, as the visibility of tourists dining at each restaurant was somewhat like the public display of its owner's capability. However, only to those with frequent interaction, an individual needs to deliberately maintain a good impression of oneself. The importance of this individual's reputation diminishes with the decrease of interaction frequency (Bailey 1971). As the restaurant owners were uncertain about the revisits of their tourist customers, they tended to look at the current transaction as a single incident, and were less reluctant to lower their meal quality to reduce costs.

A closed-system worldview, on one hand, could promote envy and jealousy among individuals, as emphasized by Foster (1965). Based on Nash's research (2007: 37) on the rise of the violence incidences in Chiapas, Mexico, "anyone who demonstrated even a modicum of wealth evoked intense envy. It was assumed that the wealthy must have gained their cattle or stores of grain at the expense of their neighbors." During her fieldwork in Zambia, Cliggett (2015) found envy and jealousy in action: the pesticide poisoning (and murder) of an ambitious man pursuing education as a path to a better life; the burning of another man's chicken coop just as his efforts to start a small-scale business began to show a return on his hard work; and numerous cases of witchcraft accusations. Cliggett suspected that her examples were not unique among the anthropologists working in impoverished places.

A closed-system worldview, on the other hand, could sustain strong leadership and lead to widespread cooperation, as spelled out by Nash (1994,

2007) and Trawick and Hornborg (2015: 13). Dipert (2001: 46–51) points out that there are two types of cooperative arrangements, unformalized and formalized. He thinks that cooperative arrangements should generally be informal rather than formal, and that there is a huge array of nonformalized cooperative behaviors. As Tucker (2010: 934) states, "competing to remain equal" might be especially intense in a small destination because of the high visibility, and therefore it is easy for any restaurateur to see the menus, décor, and service of the other restaurants. High visibility likely motivates cooperation: with higher openness in smaller places, individual behaviors are more observable, which produces binding effects, as publicly observable behaviors have a clear impact on others. Compliance is most widely achieved by informal social agreements and their associated vague societal pressures—with the frown, the stare, the remark, stigmatizing deviants' children, as further explained by Nash (2007).

SOLUTION TO THE "TRAGEDY OF THE COMMONS"

Hardin (1968) proposed two solutions to avoid the tragic outcome in "the tragedy of the commons": 1) coerced into doing it (for individuals to exert a form of mutual self-restraint) by an authoritarian state; and 2) having the resource privatized so that it can be allocated in a more selfish and competitive way by markets. These two solutions were "either socialism or the privatism of free enterprise" (Ostrom et al. 1999: 278). Trawick and Hornborg (2015: 13) believe that at a relative small scale and on a local level, people can avoid "the tragedy of the commons" on their own without a supporting institutional context provided by outsiders. Thus, there exists a third solution: the rationality of internal "cooperation" among the herdsmen. In order to prevent price wars and other disorderly competition, the restaurants in Longcun tried to negotiate a unified meal price among themselves. Even though it was not well implemented, such collective agreements are achievable and sustainable, since people could find out easily whether the rules are generally being obeyed. The rental van business in Longcun is a successful example of such.

The rental van was called "*zhongzhuang che*" (transfer vehicle) in Longcun. It refers to the MiniPassenger vans bought by some villagers to transfer the *san ke* between the village and the market town of Shanjiang, where no public transportation was available. The *san ke* did not participate in tour groups, thus there was no chartered tour shuttle service for them. When San Ge was doing his own rental van business in 2011, Lu was in charge of Longcun's tourism. Lu's company had two mini-buses to transport the *san ke*, leaving the village van drivers like San Ge very few customers. Since the merger, Lu's company withdrew, and there were no mini-buses. And the village van drivers did not allow travel agencies' tour buses to take the *san ke*, making their vehicles almost the only option for the *san ke*. The rental vans increased from a couple to a dozen in 2013.

During summer peak season, each van could get commissioned for as many as ten trips. One day, I accompanied San Ge to take a group of *san ke* to Shanjiang, where they parked their sedan. After dropping them, it was close to noon. I knew San Ge had not had a chance to eat breakfast since getting up at six o'clock, so I insisted on having some rice noodles before heading back to the village. We ordered two bowls of rice noodles. I noticed San Ge added some cold water into his hot noodle soup before eating, so that he could finish it quickly. He was too anxious to take his time to enjoy it hot. As there seemed many *san ke* in Longcun that day, it wouldn't take long before it was his turn again.

The van drivers lined up their vans in the village parking lot every morning, waiting for their turn to take customers. If one missed his turn, he became the last of that round. And the next round followed the new rotation. For a one-way trip, it was 50 yuan, which could be shared by all the passengers. These van drivers usually offered their fellow villagers a free ride if it was on their way; otherwise, they charged them a discounted price of 30 yuan. Right before the summer of 2013, the van drivers discussed and decided together to increase the price from 50 to 60 yuan. The collective agreement among the drivers ensured the fair allocation of resources, and effectively prevented price wars. Compared to running a family restaurant, the rental van operation was simpler, due to: 1) the operators interacted with their tourist customers directly, without tour guides in between; 2) the costs were almost identical, with few means to lower the costs; and 3) higher visibility led to greater cooperation.

GETTING RICH "GRADUALLY AND COLLECTIVELY"

The "limited good" and its associated "competing to remain equal" worldview of traditional peasant society contrasts with the worldview of the globalized society (Table 7.1). The juxtaposition of the two worldviews helps better understand the rationality behind economic behaviors of peasant entrepreneurs. Their conservativeness should not be simply attributed to the "backwardness." Just as Li (2008) argues, the distinction between "moral economy" (Scott 1976) and "rational peasants" (Popkin 1979) does not lie in whether it is rational, but in what kind of rationality (survival or economic). Survival rationality does not produce overall profit maximization, hence it is

Table 7.1.   Comparison between the traditional peasant society and the globalized society worldviews.

Traditional Peasant Society	Globalized Society
A closed-system	An open-system
The "limited good"	The "unlimited good"
Competing to remain equal	Competing to get ahead
Survival rationality (In acquaintance society, "face" matters most.)	Economic rationality (In commodity society, profit takes priority.)

not economic rationality, but it is sill rationality for survival. Survival rationality is not based on efficiency, but on justice and rights. The official discourse regarding development promotes individuals getting rich first through "competing to get ahead"; in contrast, the "limited good" peasant viewpoint advocates collective affluence through "competing to remain equal."

The case of Longcun has broader implications. It is not a stand-alone case. It resembles the situation in other Miao villages in China. The villagers in Langde said, "We want reform, but we would like to find our own way based on our particular reality; we desire affluence, but we expect to gain wealth gradually and collectively; we don't want become rich suddenly but steadily" (Li and Wang 2004). Not only the Miao villagers but also other ethnic minority peasants who participate in rural tourism in China have similar appeal. Based on her research in the Dai community (in Xishuangbanna of Yunnan Province) and the Zhuang community (in Yangshuo of Guangxi Province), Sun (2006) summarized the state of the local villagers' participation in tourism deveopment as high enthusiasm, low participation degree, and prevalent conflicts. She (2006: 61) points out that those villagers involved in rural tourism tended to "request to gain equal economic benefits at the moment."

Fei and Chang state, "In a village where the farms are small and wealth is accumulated slowly, there are very few ways for a landless man to become a landowner, or for a petty owner to become a large landowner. . . . It is not going too far to say that in agriculture there is no way really to get ahead. . . . To become rich one must leave agriculture" (1945: 227). In many ethnic regions in China, migrant wage laborer and tourism involvement were the two main options for those who left agriculture. In the context of tourism development, the local peasants might have adopted elements of official "development discourse," and desired to achieve material affluence. But the current development model, which is derived from the "unlimited good" and "competing to get ahead" worldview, advocates individual rather than collective prosperity, and deepened social inequality. It contradicts the traditional peasant worldview of the "limited good" and "competing to remain equal." Social pressure based on egalitarian morality (e.g., envy and jealousy) worked as an effective mechanism, to strive for not only prosperity, but also equality.

Rather than achieving prosperity for everyone, neoliberalism has produced the most skewed distribution of income and resources the world has seen (Davies et al. 2008). The increasingly fierce "competing to get ahead" continues to exacerbate social and economic differences. Inequality is socially erosive; inequality destroys trust, cohesion, and mutual aid. Reducing inequality increases the well-being and quality of life for us all (Wilkinson and Pickett 2009). It might be time to consider replacing the open-system viewpoint of the "unlimited good" promoted by the mainstream economists with the closed-system peasant worldview of the "limited good."

NOTE

1. This chapter was revised from "竞争与(不)平等：湖南凤凰苗寨游家庭餐馆的个案研究 [Competition and (In)equality: A Case Study of the Family Restaurants in a Touristic Miao Village in Fenghuang County of Hunan, China]," 旅游学刊 [*Tourism Tribune*] 31(3): 26–34.

Conclusion

"Small" as a Solution

Unhampered growth often leads to inequality and unsustainability, especially when it is elite-directed (Bodley 2003). Fenghuang's large-scale and capital-intensive tourism development illustrates how "elite-directed growth is an ill-fated process in which elite directors set off a vicious cycle of increasing scale and complexity that causes wealth and power to be concentrated in a relatively few hands and leaves the majority to pay the costs," as stated by Bodley (2013: 27). Local consequences of this global process in the case of Fenghuang, as discussed in previous chapters, include: the tourists (as well as the toured) are being objectified for profit; the government officials and private developers are reaping most of the material rewards through acquiring control of tourism resources and spatial (therefore socioeconomic) order; and the local Miao peasants are further marginalized due to the "poverty of resources," despite their everyday resistance to the unequal distribution of tourism benefits.

Similar problems have been observed elsewhere in China in other ethnic minorities (e.g., the Dong, Buyi, Dai, Zhuang) as well as Han touristic villages (Bao and Sun 2006; Chen et al. 2013; Li 2014). The villagers' involvement in tourism is uneven, and tourism benefits are unequally distributed with only a minority benefiting; the majority do not benefit, but have to share associated costs such as the shortage of goods, services, and inflation; and in some cases, due to the forceful intrusion of the government and outside capital, the villagers are marginalized and cannot participate in tourism management and decision making.

Donaldson points out, "The connection between tourism and poverty reduction depends fundamentally on whether the tourism industry is designed to include or exclude the participation of the poor" (2011: 105). Based on his research in Yunan and Guizhou Provinces of China, Donaldson (2007) concludes that tourism in Yunan has contributed to rapid economic growth but not as much in rural poverty reduction, while small-scale tourism in Guizhou has not contributed significantly to growth, but has reduced poverty by allowing poor people to participate directly.

These realities challenge the popular belief in the necessity of solving our problems by "scaling up" economic growth. As Bodley (2013: 1–2) argues, "Contemporary problems are really problems of scale and social power," and that "solutions to problems of scale and power can be readily found in successful small nations," which emphasize community solidarity and balance between all forms of wealth and between cooperation and competition, to ensure that everyone can enjoy an irreducible minimum of material well-being and personal security. Scale matters, and "small" might be a solution toward developmental justice.

The issue of scale is evident by looking at different tourism development models in rural China. Three types of tourism development models are commonly found, namely, the "outside capital-dominated model," "local government-directed model," and "community-based model" (Bao and Sun 2006; Chi and Cui 2006; Li 2014). While Fenghuang's tourism development is a good example of the outside capital-dominated model, the local government-directed model is well represented by the Xijiang Miao Village (in Guizhou), and the Langde Miao Village (also in Guizhou) represents the community-based model.

During a visit to Guizhou Minzu University in the summer of 2016, I made an unplanned overnight trip to Langde, accompanied by one of its graduate students. A backpacker we encountered in Langde, who nicknamed himself "Hiker," told us about his impressions of Fenghuang, Xijiang, and Langde, saying, "Xijiang is like the combination of Fenghuang and Langde. It has highly commercialized riverside streets which resembles Fenghuang, with mountain slope where sits the well-preserved old Miao village which resembles Langde." I recalled that some of my informants in Fenghuang, including Zhi and Hai, had been to Xijiang. Among the pictures on the wall of Zhi's living room, one was of him in Xijiang. Comparing Xijiang with tourism in his village of Longcun, Zhi was impressed with Xijiang's large scale and seeming orderliness.

Langde and Xijiang are about 12 miles apart, both located in the Qiangdongnan Miao and Dong Autonomous Prefecture in Guizhou. Belonging to one clan, Langde has 134 households with a population of 540. Xijiang is composed of different clans with 1,291 households (Chen et al. 2013: 76; Li 2014: 230), and Schein describes it as "a huge settlement of more than 5,000 people" (2000: 28). Comparing Xijiang and Langde's distinct tourism development models, Li (2014: 132–133) argues that the size of the village is a determining factor: due to its large size, Xijiang resorts to the government-directed tourism development model for its high-handed management, while Langde adopts the community-based model, which is feasible due to its small size.

The success of the community-based model relies on a self-governing consciousness, an ability to exclude outside investors, and a trustworthy leadership team (Chi and Cui 2006). In a contrastive study between China and the West examing community participation in tourism development, Bao and Sun (2006) argue that community participation in China is superficial (if not non-existent), due to the underdevelopment of democracy and non-government organizations, as well as its land tenure system in which the peasant households have contractual use rights, but not ownership rights, to farmland (as I have discussed in Fenghuang's case in Chapter 1). Ying and Zhou (2007) make a similar argument regarding the "communal" approach for tourism development in a comparative case study of two rural tourism destinations (Xidi and Hongcun) in Anhui Province of China.

The community-based model may lead to the "tragedy of the commons." The "tragedy of the commons" observed among the family restaurant businesses in Longcun of Fenghuang (in Chapter 7) is not uncommon in rural tourism destinations in China. Chi and Cui (2006: 18–19) note that in three rural tourism sites in Hangzhou of Zhejiang, such "tragedy" has been manifested as the non-regulated use of tourism resources, the chaos and disorder in competing for tourist businesses, and the lack of maintenance and reinvestment in tourism resources and facilities. How could the community-based tourism be organized in a way to prevent free riders and to promote equitable distribution of tourism benefits in a sustainable way? The Langde model might provide one possible solution.

Since its tourism inception in 1987, Langde has followed the principle that "since every villager has contributed to the construction and preservation of the village, they shall all benefit from tourism" (Chen et al. 2013: 75; Luo 2012: 166). Regarding the social structure of the traditional Miao society, a local Miao scholar in Fenghuang, who was the Director of the county Miao Studies Association, once said to me, "Among the Miao, there may be poor and rich (*you pin fu*), but no superior or inferior (*wu gui jian*)." The traditional value of egalitarianism observed among the Miao in Fenghuang remains prominent in Langde (Li 2008: 24). Li (2014: 97–99) states that egalitarianism of the Miao and the respect to their ritual leaders (e.g., shaman, head of the clan) are two important cultural elements for Langde to implement its "*gong fen*" (work points) system, a community-based tourism model which emphasizes a cooperative approach to tourism activities and the fair distribution of its associated benefits.

The principles of Langde's "*gong fen*" system include shared resources, equal opportunity, effort-based distribution, and caring for the vulnerable.

According to Li (2008: 21), one villager explained why the *"gong fen"* system was not adopted by other villages, "We have one mind, and they lack solidarity." Solidarity is both the foundation and the goal of its *"gong fen"* system. Due to the concern of causing conflicts and destroying solidarity, Langde has been ignoring the government suggestion to switch from its *"gong fen"* system to a "company run" system.

The Langde model is in sharp contrast to the popular "scaling-up" approach of tourism development in Fenghuang and elsewhere in China. In Fenghuang, outside investors have monopolized the management rights to its major tourism resources, and have constructed hotels, restaurants, and souvenir shops, taking most of the profits in return. In Langde, as observed by Donaldson (2011: 103), there has been little "development," and the most prominent developed tourism facilities include a motor road connecting the village to the county capital town, and the guest rooms with bathrooms furnished with toilets and showers, as many villagers' houses have been adapted to accommodate tourists' overnight stays.

Langde charges no entrance fee. Visitors, including both *san ke* (independent tourists) and tour groups, enter Langde freely. Upon the arrival of a tour group, the village performs an elaborate welcome ceremony at the village entrance, and then a show featuring traditional Miao music and dance of this region. The village charges the tour group for the show, with a price ranging from 800 yuan to a few thousand yuan, depending on the size of the tour group. Only those *san ke* who want to watch the show are asked to pay 30 yuan per person.

Thirty percent of the show income is reserved for the village collective fund; and the rest is distributed among those who participate, with different roles and degree of involvement being assigned corresponding with *"gong fen"* (work points) which determine an individual's share (see Li 2008 or Li 2014 for details). This is to prevent free riders and to reward those who contribute more. With no outside investors or merchants, what tourists spend on meals, lodging, and handicrafts goes directly into the pockets of the villagers who receive them. Despite the fact that Langde's overall tourism income is on the small scale, every penny stays within the village.

Similar to the Langde model, a Tibetan village of Yubeng in Yunan adopts a rotating system to equally involve the villagers in tourist businesses and to evenly distribute tourism income among those involved (Guo 2010). For example, the Yubeng villagers take turns to provide horseback riding transportation service to tourists, which is exactly what the rental van drivers in Longcun (as discussed in Chapter 7) did to eliminate disorderly competition and to evenly distribute tourist customers. Guo (2010) notes that Yubeng retains its rights to tourism management with no outside capital investment, and concluded that its community-based tourism development has enhanced community empowerment. Regarding community empowerment in the context of rural tourism development, Guo and Huang (2011) compare the Yubeng model with the outside capital dominated model in the Dai Ethnic Park in Xishuangbanna in Yunan, and find that Yubeng emphasizes equity, resulting in little growth, while the Dai Park prioritizes efficiency, scaling up its growth at the cost of fairness. This parallels the contrast between the Langde and Fenghuang. Again, scale matters, and "small" might be a solution toward developmental justice.

Represented by the Langde model, the community-based tourism development makes sure that economic benefits stay within the community rather than being extracted to the outside. In the case of Langde, the village takes a leading role in tourism management, and its villagers are the beneficiaries of tourism benefits. As perceived by the Langde villagers themselves, everyone benefits from a stable and reliable tourism income (Li 2008: 19). On one hand, the Langde model seems an ideal tourism development model that promises developmental justice; on the other hand, the government officials criticize its conservativeness and slow growth, which "emphasizes fairness but hurts efficiency" (Li 2008: 28). Langde had a head start in developing tourism in 1980s, but was quickly caught up by Xijiang, whose tourism boom was led by the county government since the 2000s (Li 2014). According to Li (2008: 12), a core member of the Langde tourism management team referred to the "*gong fen*" system as a "utopia," in the sense that it was in line with people's ideals, but meanwhile it could be easily shattered and eventually fail.

A week after I left Langde, I saw the pictures Hiker posted on WeChat (a popular Chinese social network), and realized that he was back in Langde. We had left Langde on the same day: I was heading for Fenghuang, and he was continuing to tour around Guizhou. Hiker told me that he decided to go back because the local government officials were accompanying a tourism developer to visit Langde that day, and the show in Langde was more grand than usual, performed by those selected from nearby villages. "Today's Xijiang will be Langde's future," Hiker said during our conversation on WeChat. "Guizhou is invested in promoting ethnic cultural tourism in its Qiandongnan Miao and Dong Autonomous Prefecture, and the plan is to develop Langde and several nearby villages as one scenic area." He told me, "I like Langde. As I was about to leave Guizhou, I came back here to take another look. It won't be the same soon."

A month and a half later, a colleague at Guizhou Minzu University sent me a news link, and the news reported that in order to propel this scenic area project, a forum was being held in Langde. It was hosted by the county government officials, with the attendance of invited scholars and the Chairman of Xijiang Tourism Company, who engaged discussions on how to build this Langde scenic area.

Bibliography

Abu-Lughod, Lila
1990. The Romance of Resistance: Tracing Transformations of Power Through Bedouin Women. *American Ethnologist* 17(1): 41–55.

Adas, Michael
1981. From Avoidance to Confrontation: Peasant Protest in Precolonial and Colonial Southeast Asia. *Comparative Studies in Society and History* 23(2): 217–247.
1986. From Footdragging to Flight: The Evasive History of Peasant Avoidance Protest in South and South-east Asia. Special Issue on Everyday Forms of Peasant Resistance in South-East Asia, edited by James C. Scott and Benedict J. Tria Kerkvliet. *The Journal of Peasant Studies* 13(2): 64–86.

Anonymous
2003. 凤凰厅志：乾隆, 道光, 光绪. 香港：天马图书有限公司 [Fenghuang Ting Local History: Qianlong, Daoguan, Guangxu. Hong Kong: Tian Ma Books Co. Ltd.].

Aris, Michael
1992. *Lamas, Princes, and Brigands: Joseph Rock's Photographs of the Tibetan Borderlands of China*. New York: China Institute in America.

Bailey, F. G.
1969. *Stratagems and Spoils—The Social Anthropology of Politics*. Oxford: Basil Blackwell.
1971. Gifts and Poison. In *Gifts and Poison—The Politics of Reputation*, edited by F. G. Bailey, pp. 1–26. Oxford: Basil Blackwell.

Bao, Jigang and Jiuxia Sun
2006. 社区参与旅游发展的中西差异 [A Contrastive Study on the Difference in Community Participation in Tourism Between China and the West]. 地理学报 [*Acta Geographica Sinica*] 61(4): 401–413.

Beck, Ulrich and Elisabeth Beck-Gernscheim
2002. *Individualization: Institutionalized Individualism and its Social and Political Consequences*. London: Sage Publications.

Beneria, Lourdes and Shelly Feldman eds.
1992. *Unequal Burden: Economic Crises, Persistent Poverty, and Women's Work*. Boulder: Westview Press.

Blumenfield, Tami and Helaine Silverman eds.
2013. *Cultural Heritage Politics in China*. New York: Springer.

Bodley, John
1999. Socioeconomic Growth, Culture Scale, and Household Well-Being: A Test of the Power-Elite Hypothesis. *Current Anthropology* 40(5): 595–602.
2001. Growth, Scale, and Power in Washington State. *Human Organization* 60(4): 367–379.
2003. *The Power of Scale: A Global History Approach*. New York: M. E. Sharpe.
2008. *Victims of Progress*. Lanham, MD: AltaMira Press.
2013. *The Small Nation Solution: How the World's Smallest Nations Can Solve the World's Biggest Problems*. Lanham, MD: AltaMira Press.

Brandtstädter, Susanne
2006. Book Review of Rightful Resistance in Rural China. *The Journal of Peasant Studies* 33(4): 710–712.

Brown, Michael
1996. On Resisting Resistance. *American Anthropologist* 98(4): 729–735.

Bryman, Alan
1998. Theme Parks and McDonaldization. In *Resisting McDonaldization*, edited by Barry Smart, pp. 101–115. London: Sage Publications.
1999. The Disneyization of Society. *The Sociological Review* 47(1): 25–47.

CACP and PGFC (Chinese Academy of City Planning and the People's Government of Fenghuang County)
2005. *Systematical Plan of the Urbanization of Fenghuang County (2002–2020)* [凤凰县城镇体系规划 (2002–2020)]. Locally accessed archives.

Cartier, Carolyn
2002. Origins and Evolution of a Geographical Idea: The Macroregion in China. *Modern China* 28(1): 79–142.

Castells, Manuel
1983. *The City and the Grassroots: A Cross-Cultural Theory of Urban Social Movements*. Berkeley: University of California Press.

Chambers, Erve
1997. Introduction: Tourism's Mediators. In *Tourism and Culture: An Applied Perspective*, edited by Erve Chambers, pp. 1–12. Albany: State University of New York.

Chang, Kyung-Sup
1993. The Peasant Family in Transition from Maoist to Lewisian Rural Industrialization. *Journal of Development Studies* 29(2): 220–244.

Chatterton, Paul and Robert Hollands
2003. *Urban Nightscapes: Youth Cultures, Pleasure Spaces and Corporate Power.* New York: Routledge.

Chayanov, Alexander (Zhenghong Xiao, trans.)
1996. *Peasant Farm Organization.* Beijing: Central Compilation and Translation Press [农民经济组织. 北京: 中央编译出版社].

Chen, Zhiyong and Zhiguo Kuang
2009. The Dilemma of Individual Rationality and Collective Action in the Community-Centered Tourism Development in Lang-de Miao Village. *Academic Exploration* 2009(3): 72–79 [朗德苗寨社区主导旅游发展中的个人理性与集体行动的困境. 学术探索 2009(3): 72–79].

Chen, Zhiyong, Lejing Li, and Tianyi Li
2013. 郎德苗寨社区旅游：组织演进、制度建构及其增权意义 [The Organizational Evolution, Systematic Construction and Empowerment Significance of Langde's Community Tourism]. 旅游学刊 [*Tourism Tribune*] 28(6): 75–86.

Cheong, So-Min and Marc Miller
2000. Power and Tourism: A Foucauldian Observation. *Annals of Tourism Research* 27(2): 371–390.

Chi, Jing and Fengjun Cui
2006. 乡村旅游地发展过程中的"公地悲剧"研究—以杭州梅家坞、龙坞茶村、山沟沟景区为例 [A Study on the "The Tragedy of Commons" in the Process of the Development of Rural Tourism Destinations in China: The Cases of Meijiawu Village, Longwu Village, and Shangougou Scenic Spot in Hangzhou of Zhejiang Province]. 旅游学刊 [*Tourism Tribune*] 21(7): 17–23.

Chio, Jenny
2011. Leave the Fields without Leaving the Countryside: Modernity and Mobility in Rural, Ethnic China. *Identities: Global Studies in Culture and Power* 18: 551–575.
2014. *A Landscape of Travel: The Work of Tourism in Rural Ethnic China.* Seattle: University of Washington Press.

Cho, Mun Young
2012. "Dividing the Poor": State Governance of Differential Impoverishment in Northeast China. *American Ethnologist* 39(1): 187–200.

Chomsky, Noam
1999. *Profit Over People: Neoliberalism and Global Order.* New York: Seven Stories Press.

Chu, Yin-wah and Alvin Y. So
2010. State Neoliberalism: The Chinese Road to Capitalism. In *Chinese Capitalisms: Historical Emergence and Political Implications*, edited by Yin-wah Chu, pp. 46–72. Basingstoke, UK: Palgrave Macmillan.

Church, Andrew and Tim Coles, eds.
2007. *Tourism, Power, and Space*. London: Routledge.

Cliggett, Lisa
2015. Comments. In Revisiting the Image of Limited Good: On Sustainability, Thermodynamics, and the Illusion of Creating Wealth, by Paul Trawick and Alf Hornborg, p. 17, *Current Anthropology* 56(1): 1–27.

Cockcroft, James
1998. Gendered Class Analysis: Internationalizing, Feminizing, and Latinizing Labor's Struggle in the Americas. *Latin American Perspectives* 25(6): 42–46.

Cohen, Myron L.
1993. Cultural and Political Inventions in Modern China: The Case of the Chinese "Peasant." *Daedalus* 122(2): 151–170.

Colburn, Forrest, ed.
1989a. *Everyday Forms of Peasant Resistance*. Armonk, NY: M. E. Sharpe, Inc.

Colburn, Forrest
1989b. Introduction. In *Everyday Forms of Peasant Resistance*, edited by F. D. Colburn, pp. ix–xv. Armonk, NY: M. E. Sharpe, Inc.

Cone, Abbott
1995. Crafting Selves: The Lives of Two Mayan Women. *Annals of Tourism Research* 22(2): 313–327.

Cook, Sarah
1998. Work, Wealth, and Power in Agriculture: Do Political Connections Affect the Return to Household Labor? In *Zouping in Transition: The Process of Reform in Rural North China*, edited by Andrew Walder, pp. 157–183. Cambridge, MA: Harvard University Press.

Croll, Elisabeth and Huang Ping
1997. Migration for and against Agriculture in Eight Chinese Villages. *The China Quarterly* 149: 128–146.

Davies, James, Susanna Sandstrom, Anthony Shorrocks, and Edward Wolff
2008. The World Distribution of Household Wealth. *World Institute for Development Economics Research Discussion Paper* 2008/03. Helsinki: World Institute for Development Economics Research.

Davin, Delia
1995. Women, Work and Property in the Chinese Peasant Household of the 1980s. In *Male Bias in the Development Process* (second edition), edited by Diane Elson, pp. 29–50. Manchester and New York: Manchester University Press.

Diamond, Norma
1988. The Miao and Poison: Interactions on China's Southwest Frontier. *Ethnology* 27(1): 1–25.
1995. Defining the Miao: Ming, Qing, and Contemporary Views. In *Cultural Encounters on China's Ethnic Frontiers*, edited by S. Harrell, pp. 92–116. Seattle: University of Washington Press.

Dipert, Randall
2001. Sidestepping the Tragedy of the Commons. In *The Commons: Its Tragedies and Other Follies*, edited by Tibor Machan, pp. 27–58. Stanford: Hoover Institution Press.

Di Giovine, Michael A.
2009. *The Heritage-scape: UNESCO, World Heritage, and Tourism*. Lanham: Lexington Books.

Donaldson, John A.
2007. Tourism, Development and Poverty Reduction in Guizhou and Yunnan. *The China Quarterly* 190: 333–351.
2011. *Small Works: Poverty and Economic Development in Southwestern China*. Ithaca: Cornell University Press.

Du, Shanshan
2000. "Husband and Wife Do It Together": Sex/Gender Allocation of Labor among the Qhawqhat Lahu of Lancang, Southwest China. *American Anthropologist* 102(3): 520–537.

Edensor, Tim and Julian Holloway
2008. Rhythmanalysing the Coach Tour: The Ring of Kerry, Ireland. *Transactions of the Institute of British Geographies* 33: 483–501.

EDRCCASS (Environment and Development Research Center of the Chinese Academy of Social Sciences)
1999. Analysis of the Tradeoffs on Marketing National Tourism Resources in China. *Quantitative Technology of Economy Research* 10: 3–25.

Elden, Stuart
2004. *Understanding Henri Lefebvre: Theory and the Possible*. London: Continuum.

Elson, Diane
1995. Male Bias in the Development Process: An Overview. In *Male Bias in the Development Process* (second edition), edited by Diane Elson, pp. 1–28. Manchester: Manchester University Press.

Escobar, Arturo
1992. Culture, Practice, and Politics: Anthropology and the Study of Social Movement. *Critique of Anthropology* 12(4): 395–432.

Fei, Hsiao-Tung
1939. *Peasant Life in China: A Field Study of Country Life in the Yangtze Valley.*
London: Routledge.

Fei, Hsiao-Tung and Chih-I Chang
1945. *Earthbound China: A Study of Rural Economy in Yunnan.* Chicago: University
of Chicago Press.

Feng, Xianghong
2007a. Comparison of Two Village Tourism Development Models in Fenghuang
County, China. *Journal of Northwest Anthropology* 41(2): 205–216.
2007b. Gender and Hmong Women's Handicrafts in Fenghuang's "Tourism Great
Leap Forward," China. *Anthropology of Work Review* 28 (3): 17–26.
2008a. *Economic and Socio-Cultural Impacts of Tourism Development in Fenghuang
County, China.* Ph.D. Dissertation, Department of Anthropology, Washington
State University, Pullman.
2008b. Who Benefits?: Tourism Development in Fenghuang County, China. *Human
Organization* 67 (2): 207–220.

Fenghuang County Ethnic Groups Gazetteer Writing Committee
1997. *Fenghuang Xian Minzu Zhi* [*Gazetteer of Fenghuang County's Ethnic Groups*].
Beijing: China City Press.

Fenghuang County Statistics Bureau ed. (Internal Document)
Fenghuang Statistical Yearbook 2001 [凤凰县统计局编 (内部资料). 凤凰统计年
鉴. 2001].
Fenghuang Statistical Yearbook 2013 [凤凰县统计局编 (内部资料). 凤凰统计年
鉴. 2013].

Ferguson, Lucy
2011. Promoting Gender Equality and Empowering Women? Tourism and the Third
Millennium Development Goal. *Current Issues in Tourism* 14(3): 235–249.

Fletcher, Robert
2001. What Are We Fighting For? Rethinking Resistance in a Pewenche Community
in Chile. *The Journal of Peasant Studies* 28(3): 37–66.

Foster, George
1965. Peasant Society and the Image of Limited Good. *American Anthropologist*
67(2): 293–315.
1972. A Second Look at Limited Good. *Anthropological Quarterly* 45(2): 57–64.

Foucault, Michel
1975. *The Birth of the Clinic: An Archaeology of Medical Perception.* New York:
Vintage Books.
1978. *The History of Sexuality (vol.1): An introduction.* New York: Vintage Books.
1979. *Discipline and Punish: The Birth of the Prison.* New York: Vintage Books.
1984. *Des espace autres.* Architectura, Mouvement, Continuite, Oct. pp. 46–49.

Frank, Andre Gunder
1966. The Development of Underdevelopment. *Monthly Review* 18: 17–31.

Galani-Moutafi, Vasiliki
1993. From Agriculture to Tourism: Property, Labor, Gender, and Kinship in a Greek Island Village (Part One). *Journal of Modern Greek Studies* 11(2): 241–270.
1994. From Agriculture to Tourism: Property, Labor, Gender, and Kinship in a Greek Island Village (Part Two). *Journal of Modern Greek Studies* 12(1): 113–131.

Gao, Wenshu and Russell Smyth
2011. What Keeps China's Migrant Workers Going? Expectations and Happiness among China's Floating Population. *Journal of the Asia Pacific Economy* 16(2): 163–182.

Gatrell, Jay and Noga Collins-Kreiner
2006. Negotiated Space: Tourists, Pilgrims, and the Bahai Terraced Gardens in Haifa. *Geoforum* 37: 765–778.

Gaubatz, Piper
1996. *Beyond the Great Wall: Urban Form and Transformation on the Chinese Frontiers*. Stanford: Stanford University Press.

Geertz, Clifford
1973. *The Interpretation of Cultures*. New York: Basic Books.

Gentry, Kristine
2007. Belizean Women and Tourism Work: Opportunity or Impediment? *Annals of Tourism Research* 34(2): 477–496.

Giddens, Anthony
1990. *The Consequences of Modernity*. Stanford: Stanford University Press.

Gonzalez de la Rocha, Mercedes
2001. From the Resources of Poverty to the Poverty of Resources? *Latin American Perspectives* 28 (4): 72–100.
2007. The Construction of the Myth of Survival. *Development and Change* 38(1): 45–66.

Goodman, David
1983. Guizhou and the People's Republic of China: The Development of an Internal Colony. In *Internal colonialism: Essays around a Theme*, edited by D. Drakakis-Smith and S. Williams, Monograph No. 3, pp. 107–124. London: Institute of British Geographers, Developing Areas Research Group.

Gregory, Derek and John Urry eds.
1985. *Social Relations and Spatial Structures*. Houndmills: Macmillan.

Guo, Wen
2010. 乡村居民参与旅游开发的轮流制模式及社区增权效能研究－云南香格里拉雨崩社区个案 [Study on the "Alternate System Mode" Regarding Rural Residents' Participation in Tourism Development and Community Empowerment Effectiveness]. 旅游学刊 [*Tourism Tribune*] 25(3): 76–83.

Guo, Wen and Zhenfang Huang
2011. 乡村旅游开发背景下社区权能发展研究：基于对云南傣族园和雨崩社区
两种典型案例的调查 [Community Empowerment in the Context of Rural Tourism Development in China: Investigation of Two Typical Cases of the Yunan Dai
Ethnic Park and Yubeng Village]. 旅游学刊 [*Tourism Tribune*] 26(12): 83–92.

Hansen, Mette Halskov and Rune Svarverud, eds.
2010. *iChina: The Rise of the Individual in Modern Chinese Society*. Copenhagen:
NIAS.

Hardin, Garrett
1968. The Tragedy of the Commons. *Science* 162(3859): 1243–1248.

Harrell, Steven
2001. *Ways of being ETHNIC in Southwest China*. Seattle: University of Washington
Press.

Harvey, David
1989. From Managerialism to Entrepreneurialism: The Transformation in Urban
Governance in Late Capitalism. *Geografiska Annaler* B 71:3–17.
1990. *The Condition of Postmodernity: An Enquiry into the Origins of Cultural
Change*. Cambridge, MA: Blackwell.

Haynes, Douglas and Gyan Prakash, eds.
1992. *Contesting Power: Resistance and Everyday Social Relations in South Asia*.
Berkeley: University of California Press.

He, Dan and Xinzheng Wen
2013. New Entry Fees for Scenic Town. *China Daily*, http://europe.chinadaily.com
.cn/travel/2013–04/17/content_16413821.htm, accessed on May 12, 2016.

Hendrischke, Hans
1999. Provinces in Competition: Region, Identity and Cultural Construction. In *The
Political Economy of China's Provinces: Comparative and Competitive Advantage*, edited by H. Hendrischke and C. Feng, pp. 1–25. New York: Routledge.

Ho, Samuel
1995. Rural Non-Agricultural Development in Post-Reform China: Growth, Development Patterns and Issues. *Pacific Affairs* 68(3): 360–391.

Ho, Wing-Chung
2011. James Scott's Resistance/Hegemony Paradigm Reconsidered. *Acta Politica*
46(1): 43–59.

Hsing, You-tien
2008. Socialist Land Masters: The Territorial Politics of Accumulation. In *Privatizing
China: Socialism from Afar*, edited by Li Zhang and Aihwa Ong, pp. 57–70. Ithaca:
Cornell University Press.

Hu, Hsien Chin
1944. The Chinese Concepts of "Face." *American Anthropologist* 46(1): 45–64.

Huang, Haizhu
2006. The Exploration of the Various Stakeholders in Ethnic Tourism and the Non-Homonious Factors among Them—The Case of Ping-An Village in Longsheng, Guangxi. *Guangxi Social Science* 10: 68–71 [民族旅游多元利益主体非和谐因素探讨—以广西龙胜平安村为例. 广西社会科学 10: 68–71].

Jimenez, Alberto
2003. On Space as a Capacity. *The Journal of the Royal Anthropological Institute* 9(1): 137–153.

Kellner, Douglas
1995. *Media Culture: Cultural Studies, Identity and Politics Between the Modern and the Postmodern*. New York: Routledge.
1999. Theorizing/Resisting McDonalization: A Multiperspectivist Approach. In *Resisting McDonaldization*, edited by Barry Smart, pp. 186–206. Thousand Oakes: Sage Publications.

Kerkvliet, Benedict
2009. Everyday Politics in Peasant Societies (and Ours). *The Journal of Peasant Studies* 36(1): 227–243.

Khan, Azizur and Carl Riskin
2005. China's Household Income and Its Distribution, 1995 and 2002. *The China Quarterly* 182: 356–384.

Kinkley, Jeffrey
1987. *The Odyssey of Shen Congwen*. Stanford: Standford University Press.

Kinnaird, Vivian and Derek Hall
1996. Understanding Tourism Processes: A Gender-aware Framework. *Tourism Management* 17(2): 96–102.

Kinnaird, Vivian and Derek Hall, eds.
1994. *Tourism: A Gender Analysis*. Chichester: Wiley.

Kirshenblatt-Gimblett, Barbara
1998. *Destination Culture: Tourism, Museums, and Heritage*. Berkeley: University of California Press.

Kleinman, Arthur, Yunxiang Yan, Jing Jun, Sing Lee, Everett Zhang, Pan Tianshu, Wu Fei, and Jinhua Guo
2011. *Deep China: The Moral Life of the Person*. Berkeley: University of California Press.

Knight, John and Ramani Gunatilaka
2008. *Aspirations, Adaption and Subjective Well-Being of Rural-Urban Migrants in China*. Discussion Paper No. 381, Department of Economics, University of Oxford.

2010. Great Expectations? The Subjective Well-Being of Rural Urban Migrants in China. *World Development* 38(1): 114–124.

Lawrence, Denise and Setha Low
1990. The Built Environment and Spatial Form. *Annual Review of Anthropology* 19: 453–505.

Lee, Ching Kwan and Mark Selden
2006. China's Durable Inequality: Legacies of Revolution and Pitfalls of Reform. *The Asia-Pacific Journal: Japan Focus.* http://japanfocus.org/-Mark-Selden/2329.

Lee, Gary Yia and Nicholas Tapp
2010. *Culture and Customs of the Hmong*. Santa Barbara: Greenwood.

Lefebvre, Henri
1991. *The Production of Space*. Cambridge, MA: Blackwell.
2004. *Rhythmanalysis: Space, Time, and Everyday Life*. London: Continuum.

Lemoine, Jacques
1978. L'Asie Orientale. In Jean Poirier, dir. *Ethnologie Rigionale* 2. Encyclopedic de la Pleiade. Gallimard.

Leung, Alicia
2003. Feminism in Transition: Chinese Culture, Ideology and the Development of the Women's Movement in China. *Asia Pacific Journal of Management* 20(3): 359–374.

Lew, Alan
2007. Pedestrian Shopping Streets and Urban Tourism in the Restructuring of the Chinese City. In *Tourism, Power and Space*, edited by Andrew Church and Tim Coles, pp. 150–170. London: Routledge.

Lew, Alan and Alan Wong
2004. Sojourners, Gangxi and Clan Associations: Social Capital and Overseas Chinese Tourism to China. In *Tourism, Diasporas and Space*, edited by Tim Coles and Dallen Timothy, pp. 202–214. London: Routledge.

Li, Li
2008. *Morality, Rationality and Habit in "gong-fen system" of Langde Village— Field Research about Behavior Choice of Peasants*. Guizhou Normal University Master Thesis [朗德工分制中的道义，理性与惯习—农民行为选择的田野研究. 贵州师范大学硕士学位论文].

Li, Li and Xiaomei Wang
2004. Gong-fen System—The Determination and Hesitation of Langde Village. *Guizhou Daily* [李丽，王小梅. 2004. 工分制—朗德的坚守与徘徊. 贵州日报]. http://gzrb.gog.com.cn/system/2004/05/28/000619334.shtml, accessed on May 25, 2015.

Li, Lianjiang and Kevin J. O'Brien
1996. Villagers and Popular Resistance in Contemporary China. *Modern China* 22(1): 28–61.

Li, Tianyi
2014. A *Study on the Development Models of Ethnic Minority Villages in Guizhou Province*. Chengdu: Southwest Jiaotong University Press [贵州民族村寨旅游开发模式研究. 成都：西南交通大学出版社].

Li, Wei-tsu
1948. On the Cult of the Four Sacred Animals (四大门) in the Neighborhood of Peking. *Folklore* 7(1): 1–94.

Li, Yiyou
2002. 国家风景名胜区门票专营权分析 [An Analysis of Entrance Ticket Monopoly Right of National Landscape and Historic Sites]. *旅游学刊* [*Tourism Tribune*] 17(2): 39–43.

Li, Youguo
2011. A Historic Look at the Changes in the Communist Party's Political Slogan for the Past Sixty Years Since the Establishment of the New China. *Academic Forum of Nandu* 31(3): 31–34.

Little, Walter E.
2004. *Mayas in the Marketplace: Tourism, Globalization, and Cultural Identity*. Austin: University of Texas Press.

Litzinger, Ralph
2000. *Other Chinas: The Yao and the Politics of National Belonging*. Durham: Duke University Press.

Liu, Hongren and Shuting Chen
2007. Differentiation of Peasants Since the Reform and Open-up. *Journal of Shandong Agricultural Administrators' College* 23(4): 7–9 [改革开放以来的农民分化. 山东省农业管理干部学院学报 23(4): 7–9].

Luo, Fei
2012. 郎德民族旅游管理模式对社会管理创新的启示 [Ethnic Tourism in Langde Village: Its Management Model and the Implications for Innovative Social Management]. *贵州民族学院学报* (哲学社会科学版) [*Journal of Guizhou University for Nationalities* (Philosophy and Social Sciences)] 131: 165–168.

Martin, Emily
2015. *The Meaning of Money in China and the United States: The 1986 Levis Henry Morgan Lectures*. Chicago: HAU Books.

Michaud, Jean
1997. A Portrait of Cultural Resistance: The Confinement of Tourism in a Hmong Village in Thailand. In *Tourism, Ethnicity, and the State in Asian and Pacific Societies*, edited by M. Picard and R. Wood, 128–154. Honolulu: University of Hawaii.
2012. Hmong Infrapolitics: A View from Vietnam. *Ethnic and Racial Studies* 35(11): 1853–1873.

Minca, Claudio and Tim Oakes
2011. *Real Tourism: Practice, Care, and Politics in Contemporary Travel Culture.* London: Routledge.

Mohapatra, Sandeep, Scott Rozelle, and Jikun Huang
2006. Climbing the Development Ladder: Economic Development and the Evolution of Occupations in Rural China. *Journal of Development Studies* 42(6): 1023–1055.

Moore, Henrietta
1996. *Space, Text, and Gender: An Anthropological Study of the Marakwet of Kenya.* New York: Guilford Press.
2006. The Future of Gender or the End of a Brilliant Career. In *Feminist Anthropology: Past, Present, and Future*, edited by Pamela Geller and Miranda Stockett, pp. 23–42. Philadelphia: University of Pennsylvania Press.

Morais, Duarte B., Careen Yarnal, Erwei Dong, and Lorraine Dowler
2005. The Impact of Ethnic Tourism on Gender Roles: A Comparison Between the Bai and the Mosuo of Yunan Province, PRC. *Asia Pacific Journal of Tourism Research* 10(4): 361–367.

Moser, Caroline
1996. *Confronting Crisis: A Comparative Study of Household Responses to Poverty and Vulnerability in Four Poor Urban Communities.* World Bank Environmentally Sustainable Development Studies and Monographs Series no. 8.

Nash, June
1994. Global Integration and Subsistence Insecurity. *American Anthropologist* 96(1): 7–30.
2001. *Mayan Visions: The Quests for Autonomy in an Age of Globalization.* New York: Routledge.
2007. The Notion of the Limited Good and the Specter of the Unlimited Good. In *Practicing Ethnography in a Globalizing World: An Anthropological Odyssey*, pp. 35–54. Lanham, MD: AltaMira Press.

Naughton, Barry
2007. *Is the Growth of Inequality in China Over?* Paper presented at the conference on Paradigms in Flux, University of California, San Diego, April 20–22.

Nitzky, William
2013. Community Empowerment at the Periphery? Participatory Approaches to Heritage Protection in Guizhou, China. In *Cultural Heritage Politics in China*, edited by T. Blumenfield and H. Silverman, pp. 205–232. New York: Springer.

Nyiri, Pal
2006. *Scenic Spots: Chinese Tourism, the State, and Cultural Authority in China.* Seattle: University of Washington Press.

Oakes, Tim
1995a. Shen Congwen's Literary Regionalism and the Gendered Landscape of Chinese Modernity. *Geografiska Annaler* 77B (2): 93–107.

1995b. Ethnic Tourism in Guizhou: The Legacy of Internal Colonialism. In *Tourism in China: Geographical, Political, and Economic Perspectives*, edited by Alan A. Lew and Lawrence Yu, pp. 203–222. Boulder, CO: Westview Press.
1998. *Tourism and Modernity in China.* New York: Routledge.
1999. Selling Guizhou: Cultural Development in an Era of Marketisation. In *The Political Economy of China's Provinces: Comparative and Competitive Advantage*, edited by Hans Hendrischke and Feng Chongyi, pp. 31–67. New York: Routledge.
2006. The Village as Theme Park: Mimesis and Authenticity in Chinese Tourism. In *Translocal China: Linkages, Identities, and the Reimagining of Space*, edited by T. Oakes and L. Schein, pp. 166–192. New York: Routledge.

O'Brien, Kevin
2013. Rightful Resistance Revisited. *The Journal of Peasant Studies* 40(6): 1051–1062.

O'Brien, Kevin and Lianjiang Li
2006. Rightful Resistance in Rural China. New York: Cambridge University Press.

O'Hanlon, Rosalind
1988. Recovering the Subject: Subaltern Studies and Histories of Resistance in Colonial South Asia. *Modern Asian Studies* 22(1): 189–224.

Ong, Aihwa and Li Zhang
2008. Introduction: Privatizing China: Powers of the Self, Socialism from Afar. In *Privatizing China: Socialism from Afar*, edited by L. Zhang and A. Ong, pp.1–20. Ithaca, NY: Cornell University Press.

Ortner, Sherry
1995. Resistance and the Problem of Ethnographic Refusal. *Comparative Studies in Society and History* 37(1): 173–193.

Osburg, John
2013. *Anxious Wealth: Money and Morality Among China's New Rich.* Stanford: Stanford University Press.

Ostrom, Elinor, Joanna Burger, Christopher Field, Richard Norgaard, and David Policansky
1999. Revisiting the Commons: Local Lessons, Global Challenges. *Science* 284(5412): 278–282.

Polanyi, Karl
1994. *The Great Transformation: The Political and Economic Origins of Our Time.* Boston: Beacon Press.

Popkin, Samuel
1979. *The Rational Peasant: The Political Economy of Rural Society in Vietnam.* Berkeley: University of California Press.

Potter, Sulamith Heins
1983. The Position of Peasants in Modern China's Social Order. *Modern China* 9(4): 465–499.

Rack, Mary
2005. *Ethnic Distinctions, Local Meanings: Negotiating Cultural Identities in China.* London: Pluto Press.

Ritzer, George
1993. *The McDonaldization of Society.* Newbury Park: Pine Forge Press.
1998. *The McDonaldization Thesis: Explorations and Extensions.* London: Sage Publications.
2010a. Precursors: Bureaucracy and Max Weber's Theory of Rationality, Irrationality, and the Iron Cage. In *McDonaldization: The Reader*, edited by George Ritzer, pp. 26–30. Thousand Oaks, CA: Pine Forge Press.
2010b. An Introduction to McDonaldization. In *McDonaldization: The Reader* (third edition), edited by George Ritzer, pp. 3–24. Thousand Oaks, CA: Pine Forge Press.

Ritzer, George and Allan Liska
1997. "McDisneyization" and "Post-Tourism": Complementary Perspectives on Contemporary Tourism. In *Touring Cultures: Transformations of Travel and Theory*, edited by Chris Rojek and John Urry, pp. 96–112. London: Routledge.

Rofel, Lisa
1997. Rethinking Modernity: Space and Factory Discipline in China. In *Culture, Power, Place: Explorations in Critical Anthropology*, edited by A. Gupta and J. Ferguson, pp. 155–78. Durham, NC: Duke University Press.
2007. *Desiring China: Experiments in Neoliberalism, Sexuality, and Public Culture.* Durham, NC: Duke University Press.

Rozelle, Scott, Li Guo, Minggao Shen, Amelia Hughart, and John Giles
1999. Leaving China's Farms: Survey Results of New Paths and Remaining Hurdles to Rural Migration. *The China Quarterly* 158: 367–393.

Sachs, Jeffrey
2005. *The End of Poverty: Economic Possibilities for Our Time.* New York: Penguin Press.

Sahlins, Marshall
1974. *Stone Age Economics.* Chicago: Aldine Atherton, Inc.
1993. *Waiting for Foucault.* Cambridge: Prickly Pear Press.

Salazar, Noel
2005. Tourism and Glocalization: "Local" Tour Guiding. *Annals of Tourism Research* 32(3): 628–646.
2010. The Glocalisation of Heritage through Tourism: Balancing Standardisation and Differentiation. In *Heritage and Globalisation*, edited by Sophia Labadi and Colin Long, pp. 130–146. London: Routledge.

Salazar, Noel and Nelson Graburn
2016. Introduction: Toward an Anthropology of Tourism Imaginaries. In *Tourism Imaginaries: Anthropological Approaches*, edited by Noel Salazar and Nelson Graburn, pp. 1–28. New York: Berghahn.

Sargeson, Sally
2004. FULL CIRCLE? Rural Land Reforms in Globalizing China. *Critical Asian Studies* 36(4): 637–656.

Schein, Louisa
1997. Gender and Internal Orientalism in China. *Modern China* 23(1): 69–98.
2000. *Minority Rules: The Miao and the Feminine in China's Cultural Politics*. Durham: Duke University Press.

Scott, Alison MacEwen
1995. Informal Sector or Female Sector?: Gender Bias in Urban Labour Market Models. In *Male Bias in the Development Process*, edited by Diane Elson, pp. 105–132. Manchester: Manchester University Press.

Scott, James
1976. *The Moral Economy of the Peasant: Rebellion and Subsistence in Southeast Asia*. New Haven: Yale University Press.
1985. *Weapons of the Weak: Everyday Forms of Peasant Resistance*. New Haven, CT: Yale University Press.
1986a. Everyday Forms of Peasant Resistance. In Special Issue on Everyday Forms of Peasant Resistance in South-East Asia, edited by J. C. Scott and B. J. Kerkvliet. *The Journal of Peasant Studies* 13(2): 5–35.
1986b. Introduction. In Special Issue on Everyday Forms of Peasant Resistance in South-East Asia, edited by J. C. Scott and B. J. Kerkvliet. *The Journal of Peasant Studies* 13(2): 1–3.
1989. Everyday Forms of Resistance. In *Everyday Forms of Peasant Resistance*, edited by F. D. Colburn, pp. 3–33. Armonk, NY: M. E. Sharpe, Inc.
2005. Afterword to "Moral Economies, State Spaces, and Categorical Violence." *American Anthropologist* 107(3): 395–402.

Seymour, Susan
2006. Resistance. *Anthropological Theory* 6(3): 303–321.

Shen, Congwen
1934. *Bian Cheng*. Shanghai: Life Book Store [边城. 上海 : 生活书店].
1982. *Recollections of West Hunan*. Translated by Gladys Yang. Beijing: Panda Books.
1986. *Fenghuang*. Beijing: Culture and Art Publishing House [凤凰. 北京 : 文化艺术出版社].

Shepherd, Robert
2013. *Faith in Heritage: Displacement, Development, and Religious Tourism in Contemporary China*. Walnut Creek, CA: Left Coast Press.

Shi, Qigui
2002. West Hunan Miao Ethnic Group Field Survey Report. Changsha: Hunan People's Press [湘西苗族实地调查报告. 长沙：湖南人民出版社].

Silver, Ira
1999. Review Essay: Tourism, Cultural Appropriation, and Local Resistance. *Sociological Inquiry* 69(3): 504–507.

Sinclair, Thea, ed.
1997. *Gender, Work and Tourism*. New York: Routledge.

Skinner, William
1964. Marketing and Social Structure in Rural China, part I. *Journal of Asian Studies* 24(1): 3–43.
1977. *The City in Late Imperial China*. Stanford: Stanford University Press.
1985. The Structure of Chinese History. *Journal of Asian Studies* 44(2): 271–292.
1994. Differential Development in Lingnan. In *The Economic Transformation of South China—Reform and Development in the Post-Mao Ear*, edited by T. P. Lyons and V. Nee, pp. 17–54. New York: Cornell University Press.

Smart, Barry, ed.
1999. *Resisting McDonaldization*. Thousand Oakes, CA: Sage Publications.

Smith, Carol
1995. Race-Class-Gender Ideology in Guatemala: Modern and Anti-Modern Forms. *Comparative Studies in Society and History* 37(4): 723–749.

Smith, Valene
1989. *Hosts and Guests: The Anthropology of Tourism*. Philadelphia: University of Pennsylvania Press.

Song, Linfei
1996. Rural Labor Relocation and Its Strategies in China. *Sociological Studies* 2: 105–117 [中国农村劳动力的转移与对策. 社会学研究 2: 105–117].

Spencer, J. E.
1940. Kueichou: An Internal Chinese Colony. *Pacific Affairs* 13: 162–172.

Stonich, Susan, Jerrel Sorensen, and Anna Hundt
1995. Ethnicity, Class, and Gender in Tourism Development: The Case of the Bay Islands, Honduras. *Journal of Sustainable Tourism* 3(1): 1–28.

Su, Xiaobo and Peggy Teo
2008. Tourism Politics in Lijiang, China: An Analysis of State and Local Interactions in Tourism Development. *Tourism Geographies* 10: 150–168.

Sun, Jiuxia
2006. Staying on the Farmland and the Participatory Rural Community Tourism—The Participation Status and Causes of Peasants in Community Tourism. *Thinking* 32(5): 59–64 [守土与乡村社区旅游参与—农民在社区旅游中的参与状态及成因. 思想战线 32(5): 59–64].

Sun, Le
2013. "Tourist Tax" Highlights Unrestrained Power. *The Economic Observer*. http://www.eeo.com.cn/ens/2013/0503/243395.shtml, accessed on May 12, 2016.

Svensson, Marina
2010. Tourist Itineraries, Spatial Management, and Hidden Temples: The Revival of Religious Sites in a Water Town. In *Faiths on Display: Religion, Tourism, and the Chinese State*, edited by Tim Oakes and Donald Sutton, pp. 211–233. Lanham, MD: Rowman & Littlefield Publishers.

Swain, Margaret
1989. Gender Roles in Indigenous Tourism: Kuna Mola, Kuna Yala, and Cultural Survival. In *Hosts and Guests: The Anthropology of Tourism*, edited by Valene Smith, pp. 83–104. Philadelphia: University of Pennsylvania Press.
1990. Commoditizing Ethnicity in Southwest China. *Cultural Survival Quarterly* 14: 26–29.
1993. Women Producers of Ethnic Arts. *Annals of Tourism Research* 20(1): 32–51.
1995. Gender in Tourism. *Annals of Tourism Research* 22(2): 247–266.
2002. Looking South: Local Identities and Transnational Linkages in Yunan. In *Rethinking China's Provinces*, edited by John Fitzgerald, pp. 179–220. New York: Routledge.

Tang, Wing-Shing
2014. Governing by the State: A Study of the Literature on Governing Chinese Mega-Cities. In *Branding Chinese Mega-Cities: Policies, Practices and Positioning*, edited by Per Olof Berg and Emma Bjorner, pp. 42–63. Northampton, MA: Edward Elgar.

Tang, Zijun
2013. Does the Institution of Property Rights Matter for Heritage Preservation? Evidence from China. In *Cultural Heritage Politics in China*, edited by T. Blumenfield and H. Silverman, pp. 23–30. New York: Springer.

Tao, Z.
2006. *China's Migrant Workers*. Address to the National Social Administration Workshop, August, Beijing.

Tapp, Nicholas
2003. *Hmong of China: Context, Agency, and the Imaginary*. Boston: Brill Academic Publishers.

Tawney, Richard
1966. *Land and Labor in China*. Boston: Beacon Press.

Taylor, J. Edward, Scott Rozelle, and Alan de Brauw
2003. Migration and Incomes in Source Communities: A New Economics of Migration Perspective from China. *Economic Development and Cultural Change* 52(1): 75–101.

Teo, Peggy and Sandra Leong
2006. A Postcolonial Analysis of Backpacking. *Annals of Tourism Research* 33(1): 109–131.

TPCCACP and TBFC (Tourism Planning Center of Chinese Academy of City Planning and Tourism Bureau of Fenghuang County)
2004. *Tourism Development Plan of Fenghuang County*. Locally accessed archives.

Trawick, Paul and Alf Hornborg
2015. Revisiting the Image of Limited Good: On Sustainability, Thermodynamics, and the Illusion of Creating Wealth. *Current Anthropology* 56(1): 1–27.

Tucker, Hazel
2007. Undoing Shame: Tourism and Women's Work in Turkey. *Journal of Tourism and Cultural Change* 5(2): 87–105.
2010. Peasant-Entrepreneurs: A Longitudinal Ethnography. *Annals of Tourism Research* 37(4): 927–946.

Turner, Sarah
2012a. Making a Living the Hmong Way: An Actor-Oriented Livelihoods Approach to Everyday Politics and Resistance in Upland Vietnam. *Annals of the Association of American Geographers* 102(2): 403–422.
2012b. "Forever Hmong": Ethnic Minority Livelihoods and Agrarian Transition in Upland Northern Vietnam. *The Professional Geographer* 64(4): 540–553.

Turner, Sarah and Jean Michaud
2009. "Weapons of the Week": Selective Resistance and Agency among the Hmong in Northern Vietnam. In *Agrarian Angst and Rural Resistance in Contemporary Southeast Asia*, edited by D. Caouette and S. Turner, pp. 45–60. New York: Routledge.

UNWTO and UN Women (The World Tourism Organization and The United Nations Entity for Gender Equality and the Employment of Women)
2011. *The Global Report on Women in Tourism 2010*. http://www2.unwto.org/sites/all/files/pdf/folleto_globarl_report.pdf, accessed on March 24, 2012.

Van der Ploeg, Jan Douwe
2008. *The New Peasantries: Struggles for Autonomy and Sustainability in an Era of Empire and Globalization*. London: Earthscan.

Walder, Andrew and Litao Zhao
2006. Political Office and Household Wealth: Rural China in Deng Era. *The China Quarterly* 186: 357–376.

Walker, Kathy
2006. "Gangster Capitalism" and Peasant Protest in China: The Last Twenty Years. *The Journal of Peasant Studies* 33(1): 1–33.
2008. From Covert to Overt: Everyday Peasant Politics in China and the Implications for Transnational Agrarian Movements. *Journal of Agrarian Change* 8(2, 3): 462–488.

Wallerstein, Immanuel
1974. *The Modern World System*. New York: Academic Press.

Wang, Hui, Zhaoping Yang, Li Chen, Jingjing Yang, and Rui Li
2010. Minority Community Participation in Tourism: A Case of Kanas Tuva Villages in Xinjiang, China. *Tourism Management* 31(6): 759–764.

Wang, Ning
2006. Itineraries and the Tourism Experience. In *Travels in Paradox: Remapping Tourism*, edited by Claudio Minca and Tim Oakes, pp. 65–76. Lanham, MD: Rowman & Littlefield Publishers.

Watson, James ed.
1997. *Golden Arches East: McDonald's in East Asia*. Stanford: Stanford University Press.

Weaver, Adam
2005. The McDonaldization Thesis and Cruise Tourism. *Annals of Tourism Research* 32(2): 346–366.

Weber, Max
1978. *Economy and Society*, vols. 1 and 2, edited by G. Roth and C. Wittich. Berkeley: University of California Press.

White, Christine
1986. Everyday Resistance, Socialist Revolution and Rural Development: The Vietnamese Case. *The Journal of Peasant Studies* 13(2): 49–63.

Wilkinson, Richard and Kate Pickett
2009. *The Spirit Level: Why Greater Equality Makes Society Stronger*. New York: Bloomsbury.

Wilkinson, Paul and Wiwik Pratiwi
1995. Gender and Tourism in an Indonesian Village. *Annals of Tourism Research* 22(2): 283–299.

Wolf, Eric
2001. Is the "Peasantry" a Class? In *Pathways of Power: Building An Anthropology of the Modern World*, edited by Eric Wolf with Sydel Silverman, pp. 252–259. Berkeley: University of California Press.

WTO (World Tourism Organization)
2006. *Poverty Alleviation Through Tourism: A Compilation of Good Practices*. Madrid: World Tourism Organization.

Xie, Philip
2011. *Authenticating Ethnic Tourism*. Tonawanda, NY: Channel View Publications.

Xie, Yongkui and Yuanming Xu
1988. Agricultural Labor Relocation and Urbanization in China. *Journal of Renmin University of China* 1: 80–86 [我国农业劳动力转移与城市化. 中国人民大学学报 1: 80–86].

Yan, Changwen
2013. The Speech at the County Biannual Meeting of Economic Work [在全县半年经济工作会上的讲话]. Internally circulated document within the county government, transcript from Yan's oral speech on July 11, 2013.

Yan, Yunxiang
2009. *The Individualization of Chinese Society*. Oxford: Berg.

Yang, Mayfair Mei-Hui
2004. Spatial Struggles: Postcolonial Complex, State Disenchantment, and Popular Reappropriation of Space in Rural Southeast China. *The Journal of Asian Studies* 63(3): 719–755.

Yang, Xiaoxia
2004. An Analysis on a Problem of Property Rights of Tourism Resources in China. *Economic Geography* 24 (3): 416–422.

Yea, Sallie and Gabriel Noweg
2000. The Reality of Community: Women's Roles in Iban Longhouse Tourism in Sarawak. *Borneo Review* 11(2): 19–37.

Ying, Tianyu and Yongguang Zhou
2007. Community, Governments and External Capitals in China's Rural Cultural Tourism: A Comparative Study of Two Adjacent Villages. *Tourism Management* 28(1): 96–107.

Zhang, Li
2001a. Migration and Privatization of Space and Power in Late Socialist China. *American Ethnologist* 28(1): 179–205.
2001b. *Strangers in the City: Reconfigurations of Space, Power, and Social Networks within China's Floating Population*. Stanford: Stanford University Press.
2010. *In Search of Paradise: Middle-Class Living in a Chinese Metropolis*. Ithaca, NY: Cornell University Press.

Zhou, Yingying, Han Hua, and Stevan Harrell
2008. From Labour to Capital: Intra-Village Inequality in Rural China, 1988–2006. *The China Quarterly* 195: 515–534.

Zinda, John Aloysius
2015. Tourism Dynamos: Selective Commodification and Developmental Conservation in China's Protected Areas. *Geoforum*. Corrected proof available online at http://dx.doi.org/10.1016/j.geoforum.2015.08.004.

Zito, Angela
1997. *Of Body and Brush: Grand Sacrifice as Text/Performance in Eighteenth Century China*. Chicago: University of Chicago Press.

Zorn, Elayne and Linda Clare Farthing
2007. Communitarian Tourism Hosts and Mediators in Peru. *Annals of Tourism Research* 34(3): 673–689.

Zukin, Sharon
1993. *Landscapes of Power: From Detroit to Disney World.* Berkeley: University of California Press.

Zweig, David
1989. Struggling Over Land in China: Peasant Resistance after Collectivization, 1966–1986. In *Everyday Forms of Peasant Resistance,* edited by Forrest D. Colburn, pp. 151–174. Armonk, NY: M. E. Sharpe, Inc.

Index

Page references for figures and tables are italicized.

About the Author

Xianghong Feng (丰向红) is associate professor of anthropology at Eastern Michigan University. Since 2002 she has conducted fieldwork in Fenghuang County of Hunan Province in China on its evolving socioeconomic dynamic from the prism of tourism.

www.ingramcontent.com/pod-product-compliance
Lightning Source LLC
Chambersburg PA
CBHW050650280326
41932CB00015B/2849